Diese Mitteilungen setzen eine von Erich Regener begründete Reihe fort, deren Hefte auf der vorletzten Seite genannt sind.

Das Max-Planck-Institut für Aeronomie vereinigt zwei Institute, das Institut für Stratosphärenphysik und das Institut für Ionosphärenphysik.

Ein **(S)** oder **(I)** beim Titel deutet an, aus welchem Institut die Arbeit stammt.

Anschrift der beiden Institute:

                              3411 Lindau

# MEASUREMENTS OF HIGH ENERGETIC AURORAL RADIATIONS WITH BALLOON-BORNE DETECTORS IN 1962 AND 1963

compiled by G. PFOTZER and A. EHMERT

Contributions by

A. BEWERSDORFF, G. KREMSER, G. PFOTZER, L. ROSSBERG
Institut für Stratosphärenphysik
am Max-Planck-Institut für Aeronomie
3411 Lindau/Harz, Germany

W. RIEDLER
Kiruna Geophysical Observatory
Kiruna C, Sweden

H. TREFALL
Universitetet i Bergen
Fysisk Institutt
Bergen, Norway

J.P. LEGRAND
Laboratoire de Physique Cosmique
Station de Chalais-Meudon
Meudon, France

ISBN 978-3-540-03358-5   ISBN 978-3-642-87443-7 (eBook)
DOI 10.1007/978-3-642-87443-7

## Corrigendum

Captions of Fig. A1 and A2
          replace [13] by [KEPPLER, 1964]

Caption of Fig. A3
          replace [14] by [1964]

Flight K2/63
Base line of the E > 100 keV channel
please read "50/sec" instead of "25/sec".

Figures concerning technical details,

flight schedules and some special diagrams

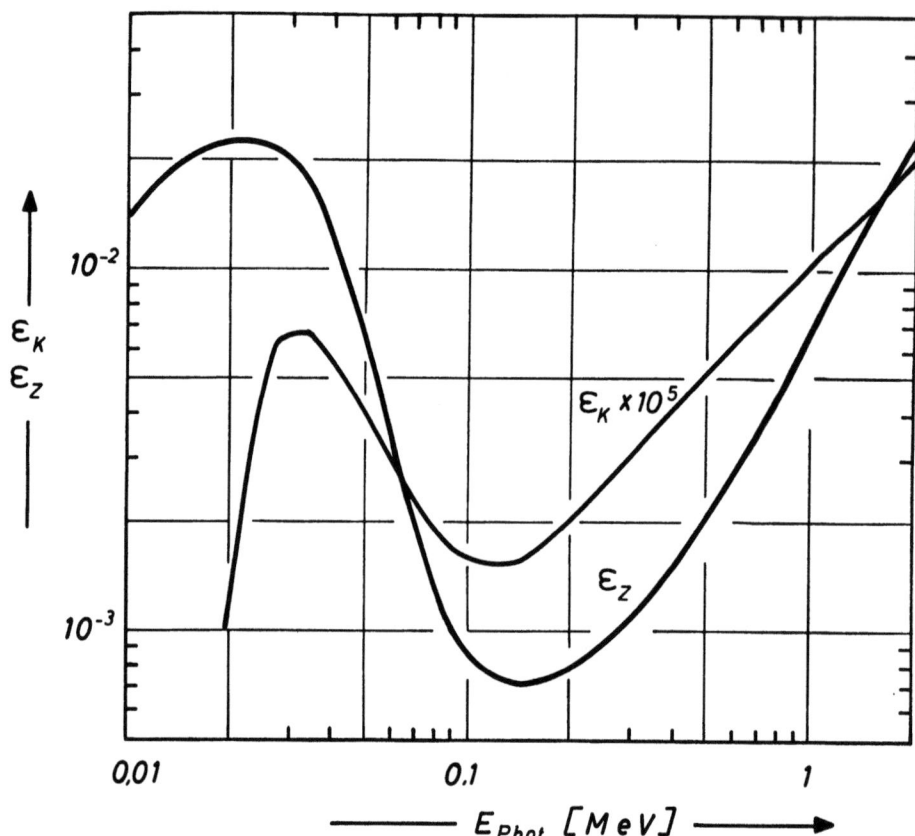

Fig. A1

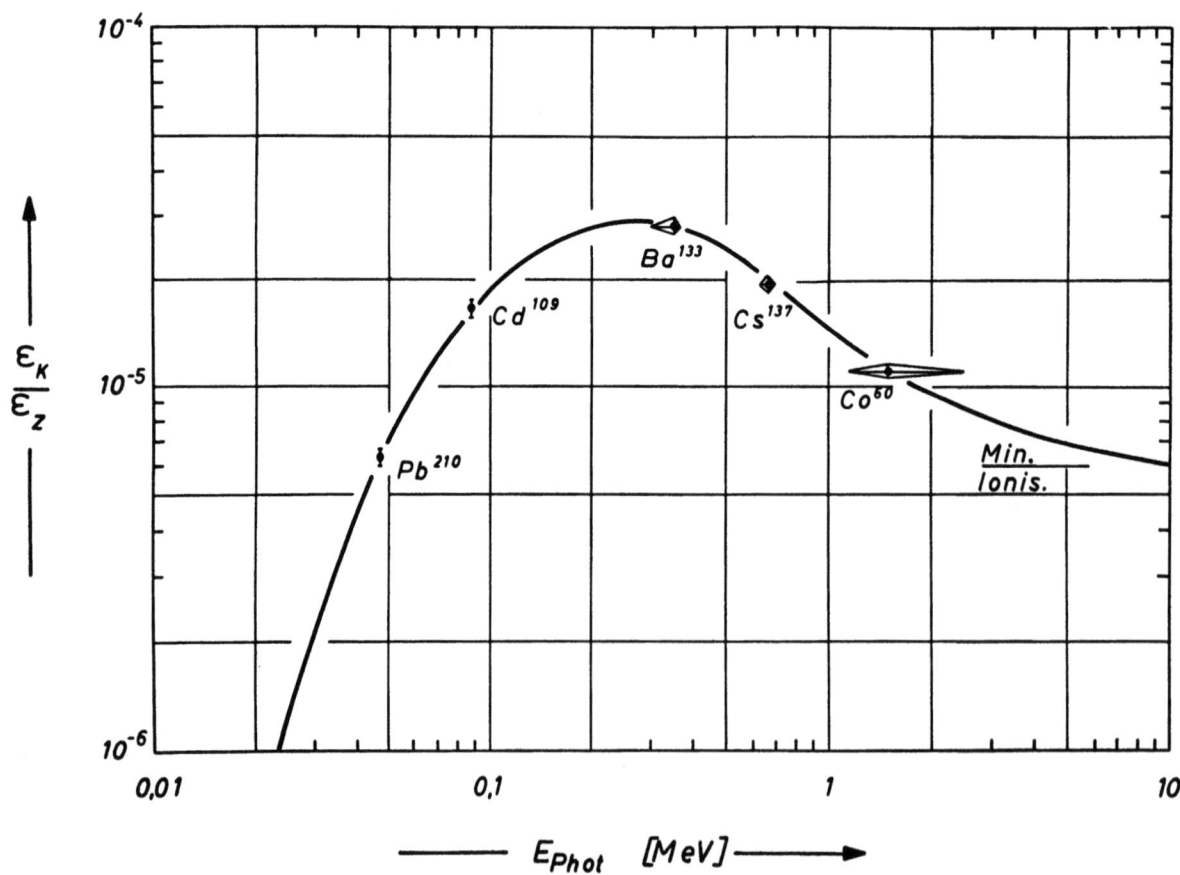

Fig. A2

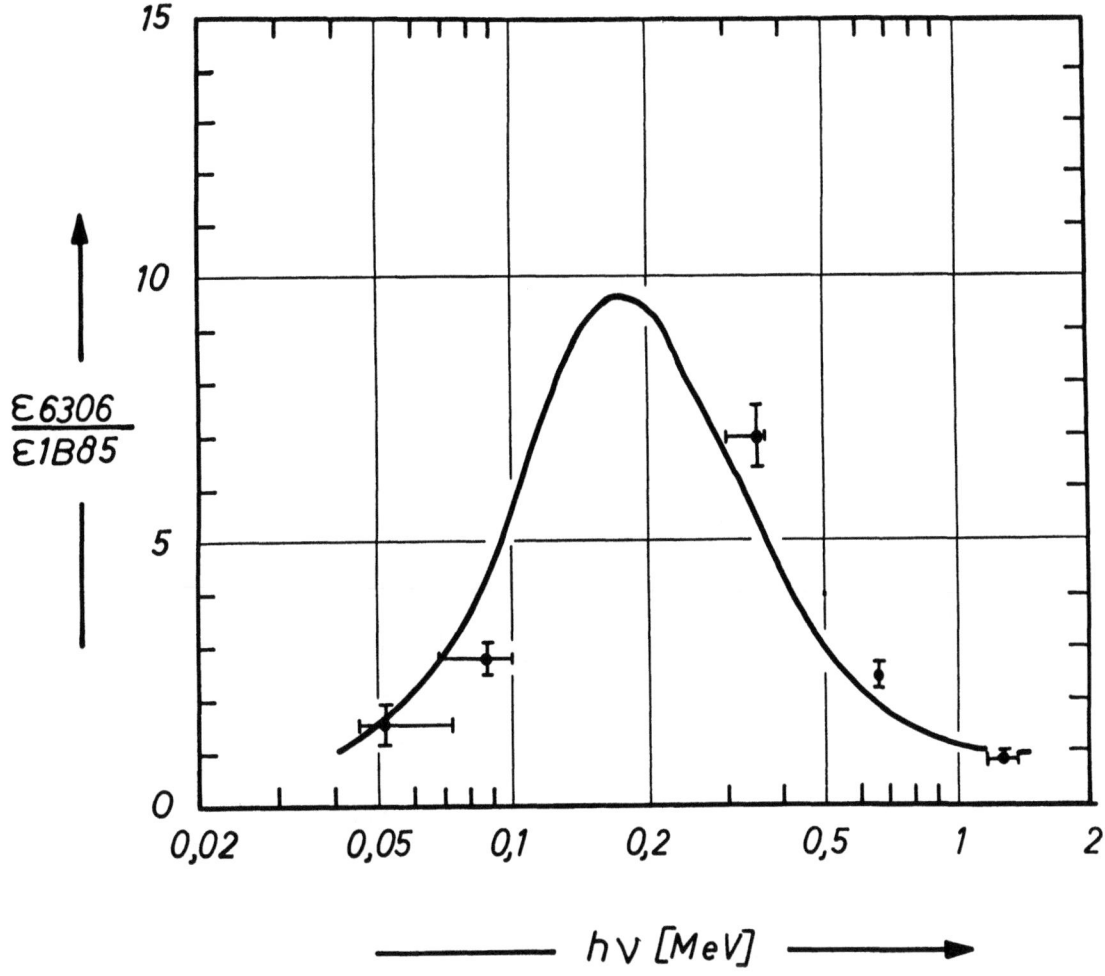

Fig. A3

Fig. A1  Efficiencies of the ionization chamber ($\varepsilon_k$) and of the 1B85 GM-counter ($\varepsilon_z$) versus the energy $E_{phot}$ of monoenergetic photons. The efficiency is defined in both cases by the equation: $N = \varepsilon\ G \cdot \Phi(E_{phot})$

    $N \;\hat{=}\;$ counting rate of the GM-counter or pulse rate of the ionization chamber

    $G \;\hat{=}\;$ respective geometry factor

    $\Phi E_{phot} \;\hat{=}\;$ flux of monoenergetic photons/(cm$^2$sec sterad)

    $\varepsilon \;\hat{=}\;$ efficiency

after [13]

Fig. A2  Ratio $\varepsilon_k/\varepsilon_z$ of the efficiencies of the ionization chamber and the 1B85 counter versus the energy $E_{phot}$ of monoenergetic photons [after 13]

Fig. A3  Ratio of the efficiencies of the GM-counters 6306 (Bi coated) and the 1B85 (Al wall only) versus the energy $h'\nu$ of monoenergetic photons. Curve according to manufacturer data modified according to direct measurements of the ratio in the telescope geometry by Richter [14]. The points ⊢ were checked with an individual set.

# Flights 1962

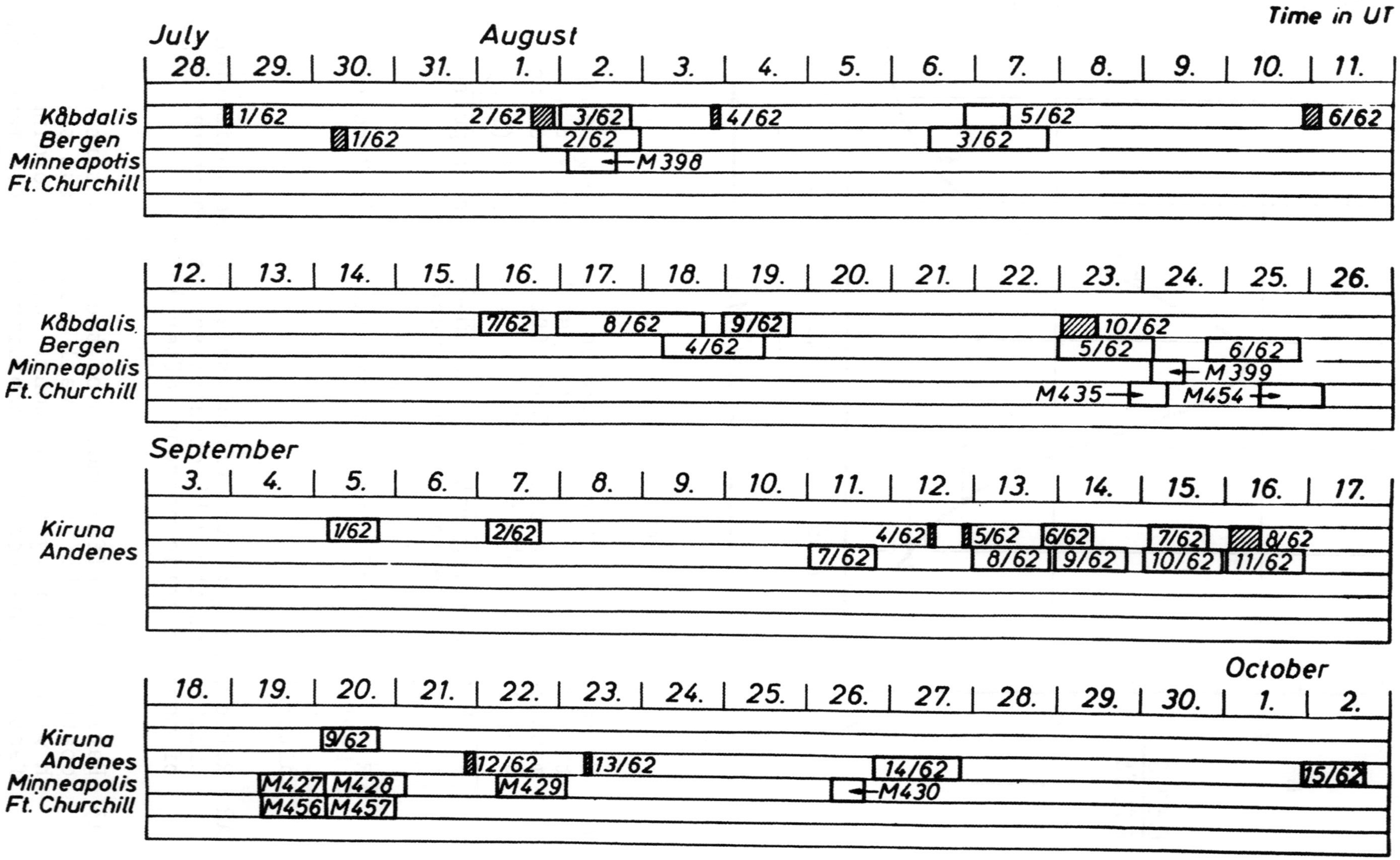

Fig. B1  Schedule of flights 1962 at different stations. Rectangular boxes indicate the duration of the flights. Boxes are hatched when the number could not be inserted inside.

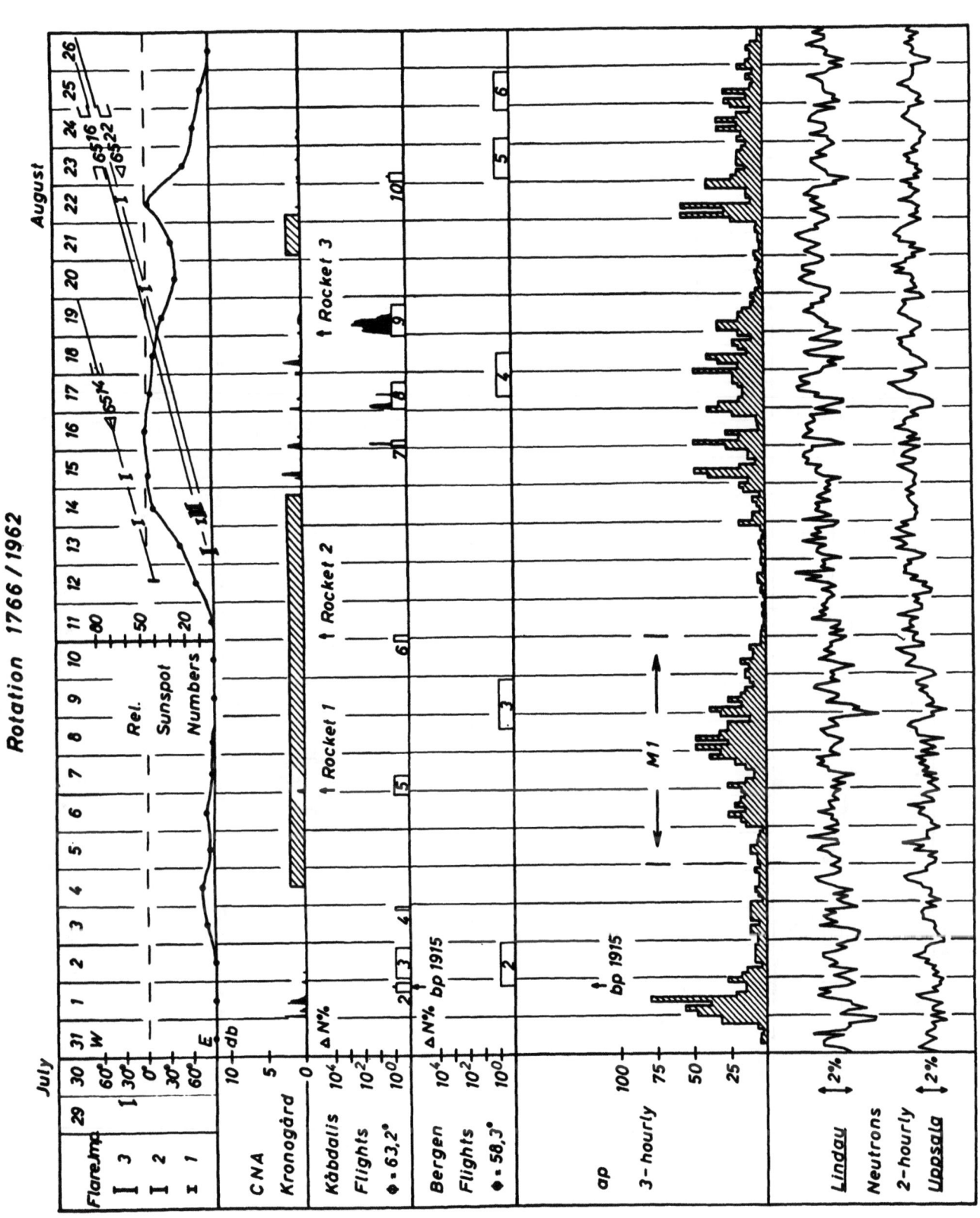

Fig. B2  Solar terrestrial relations during the solar rotation 1766

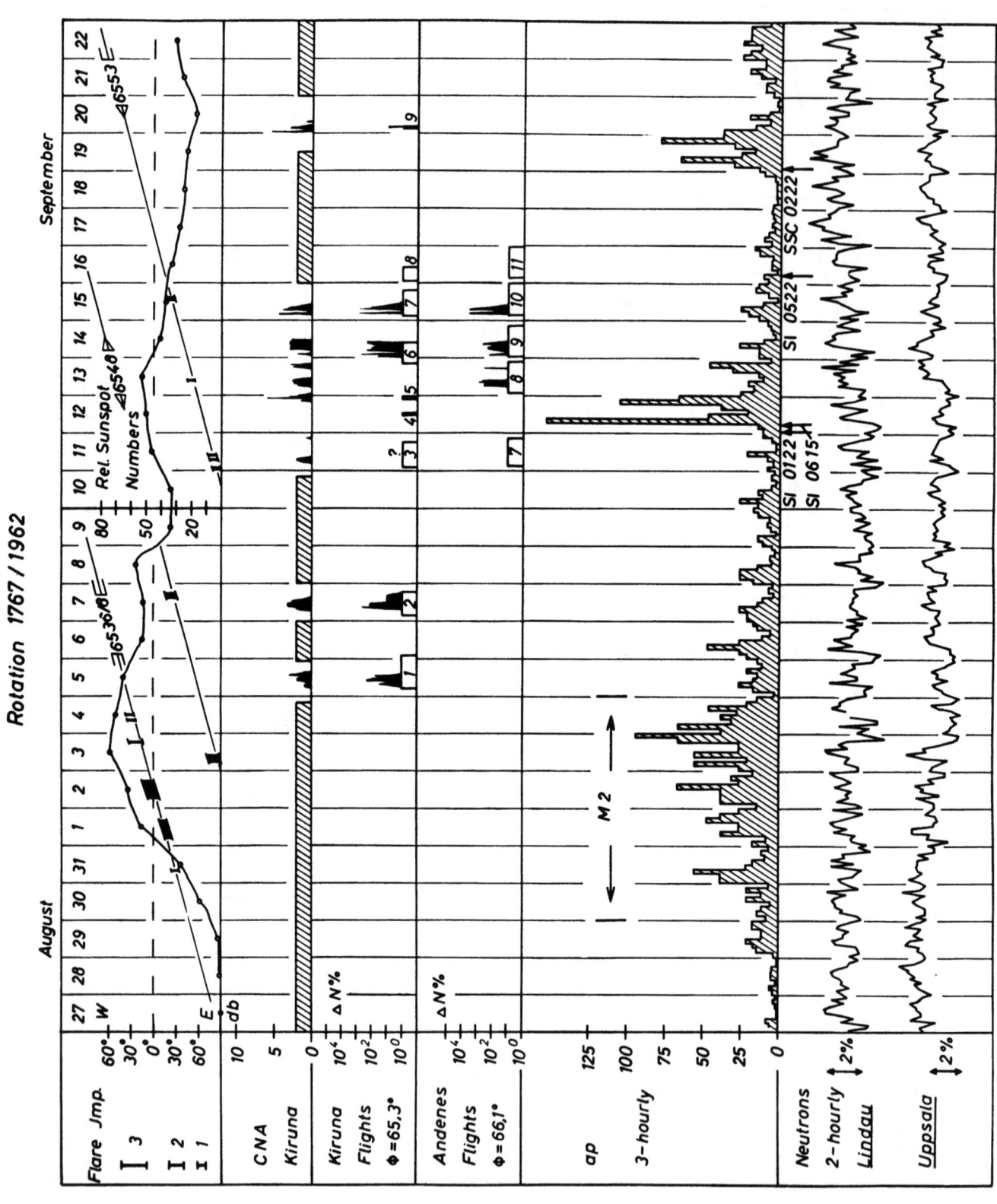

Fig. B3  Same as B2 for rotation 1767

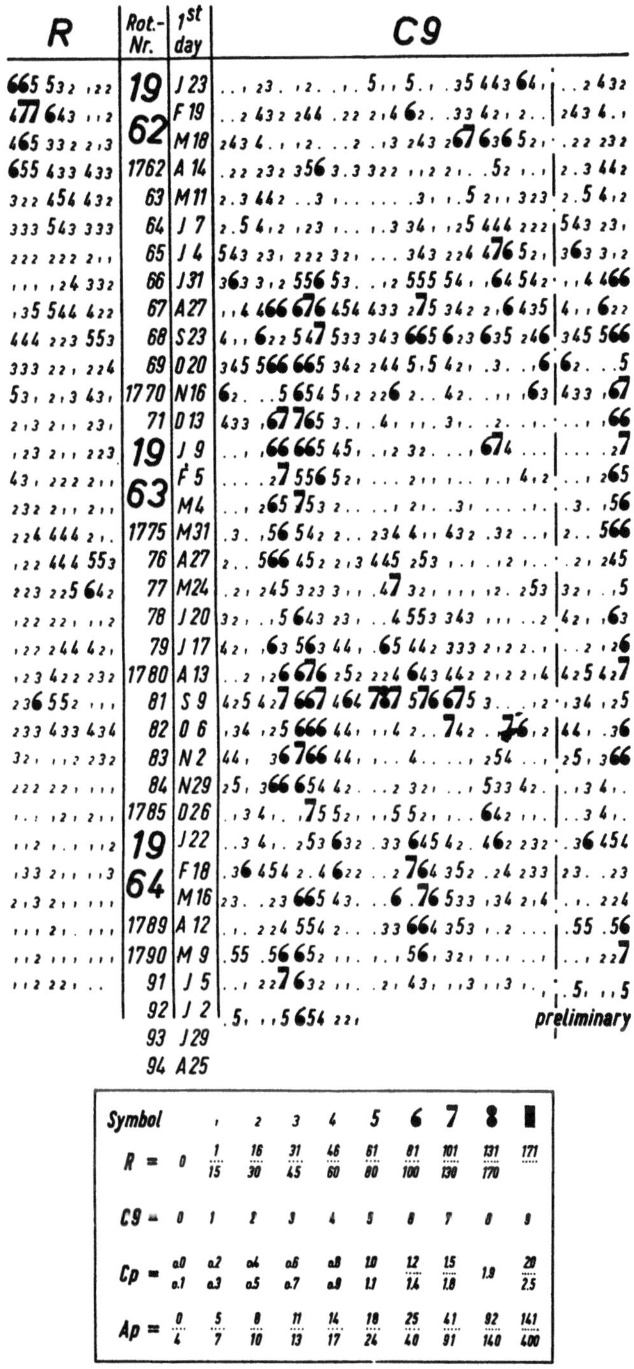

Fig. B4  Daily geomagnetic character figures C9 and sunspot numbers R
Symbolic tabulation in the 27 days sequence of the solar rotations after BARTELS. The outstanding recurrence of the M-region (dark resp. large digits) beginning at about the 6th day of every rotation is clearly seen. These diagrams are distributed by the "Geophysikalisches Institut der Universität Göttingen", on request.

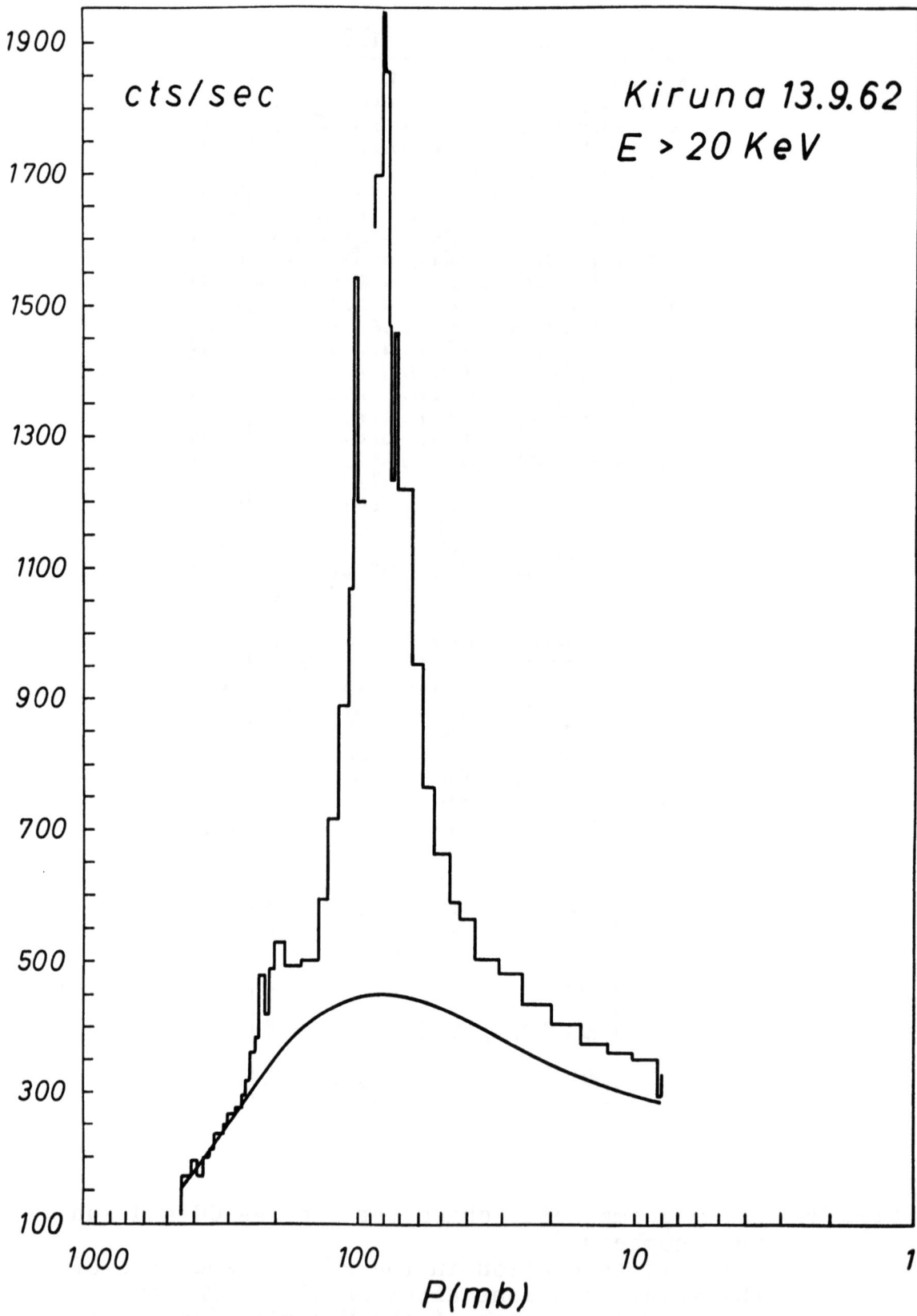

Fig. B5 Excess counting rate during flight K6/1962 versus pressure (scintillation counter)

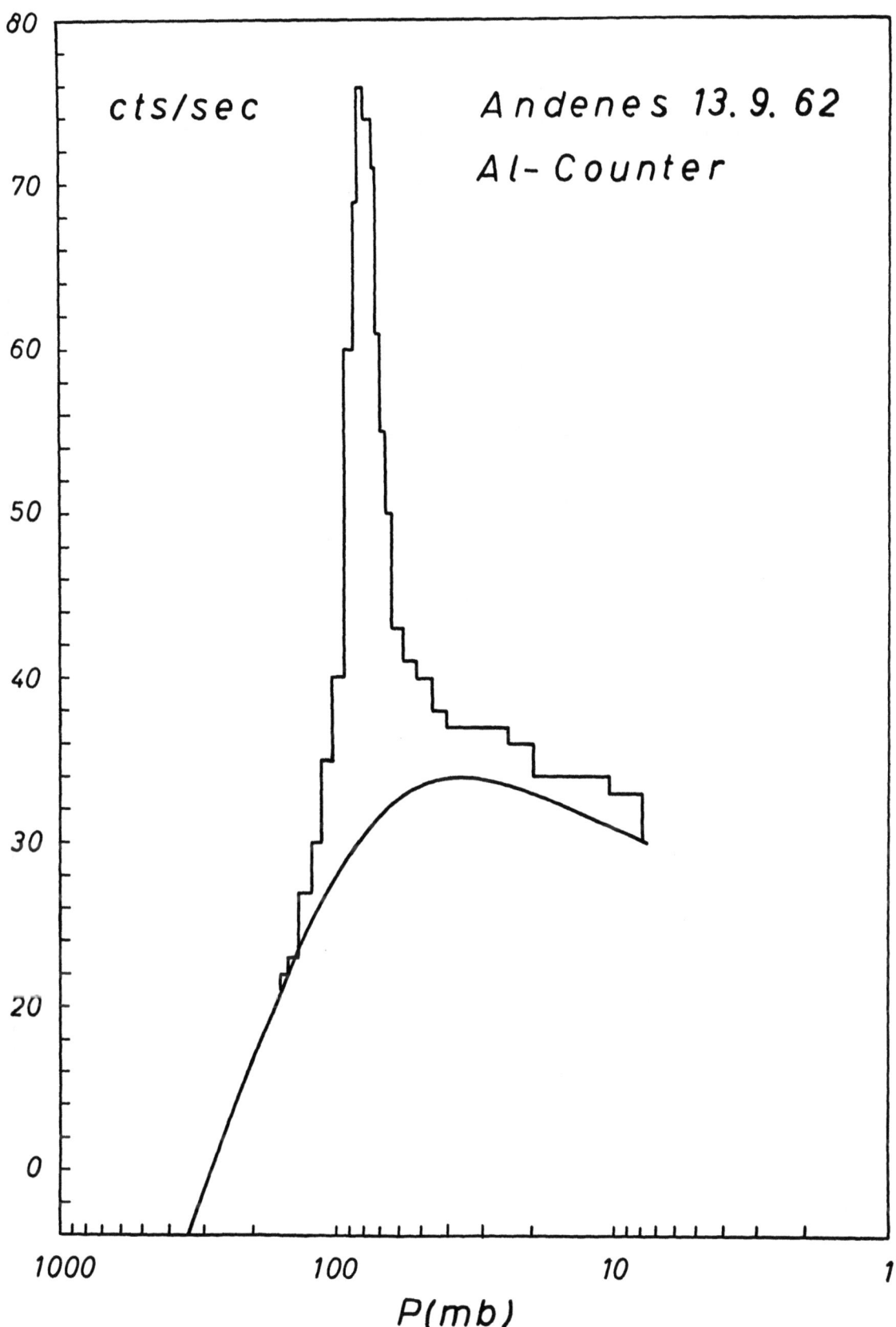

Fig. B6  Excess counting rate during flight A9/1962 versus pressure (1B85, Al-counter)

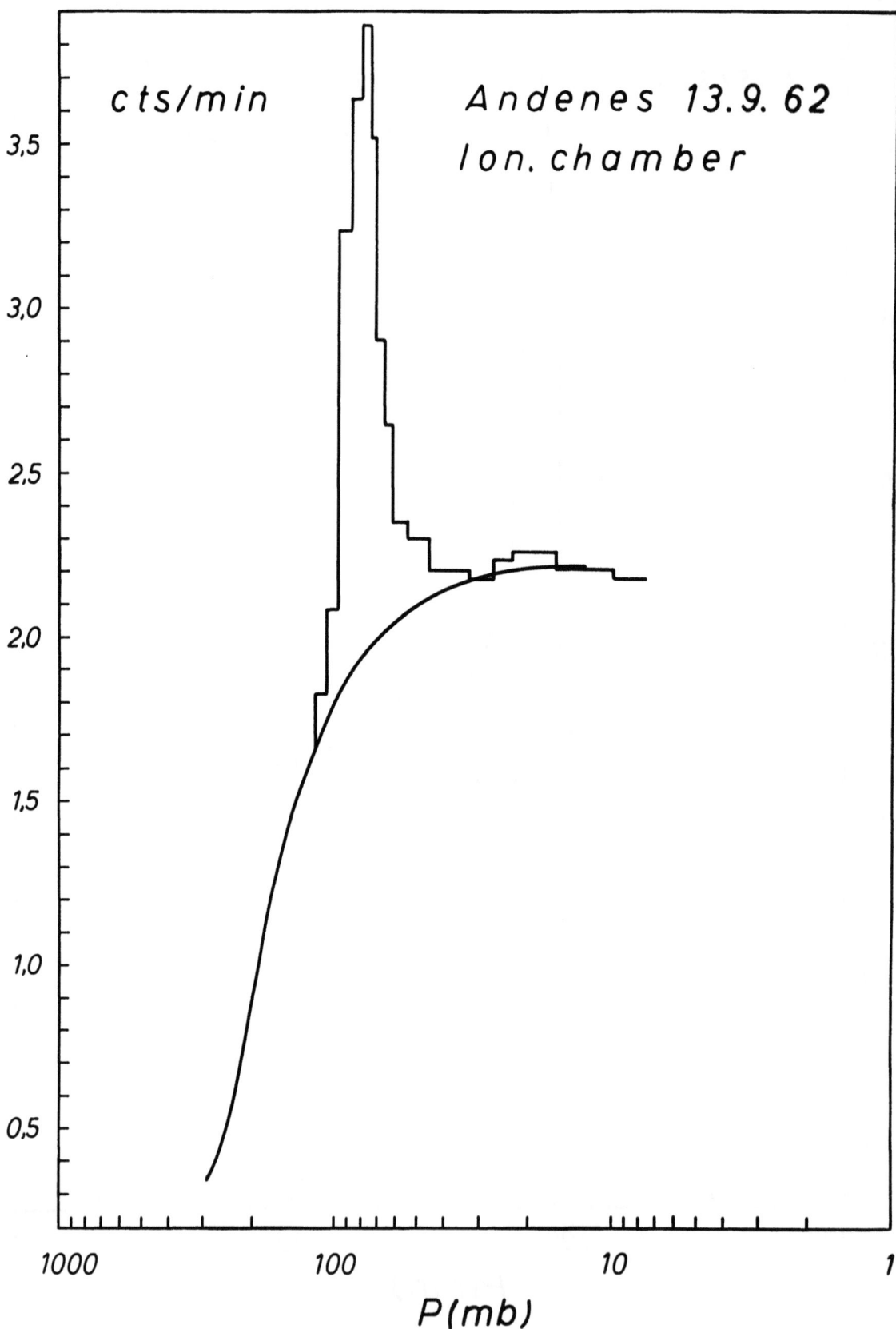

**Fig. B7**  Same as B6 (ionization chamber)

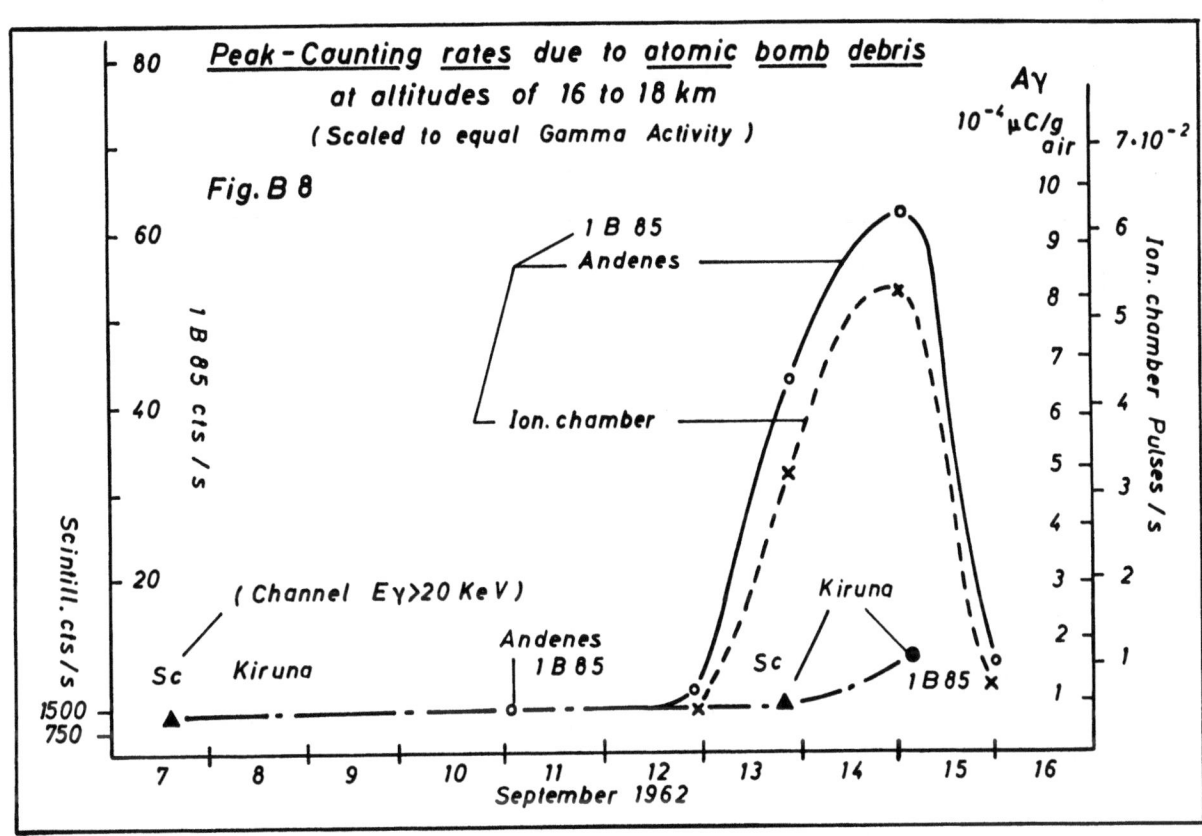

Fig. B8   Peak counting rates due to radioactive debris of various detectors during different flights over Kiruna and Andenes.

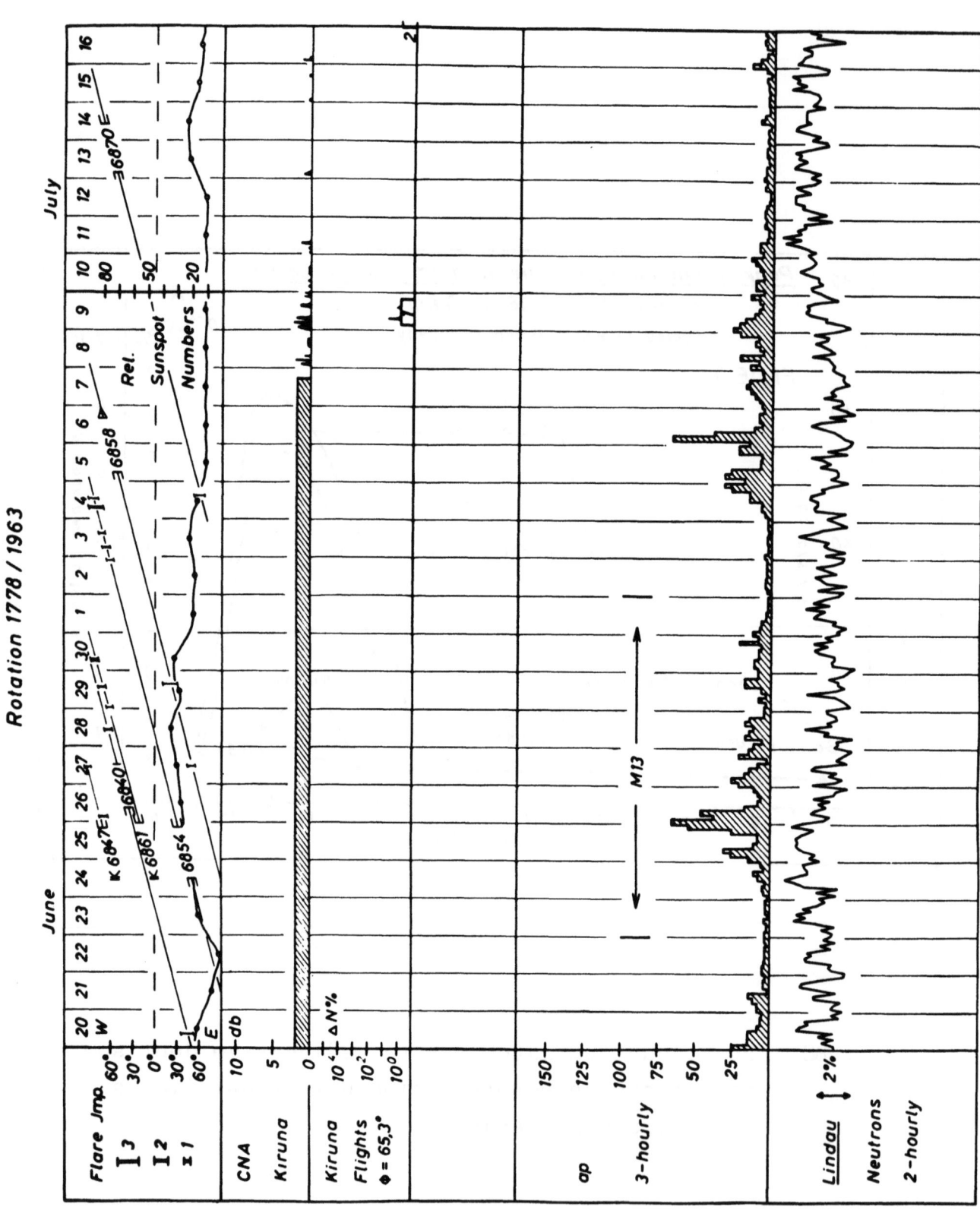

Fig. C1  Solar terrestrial relations during the solar rotation 1778

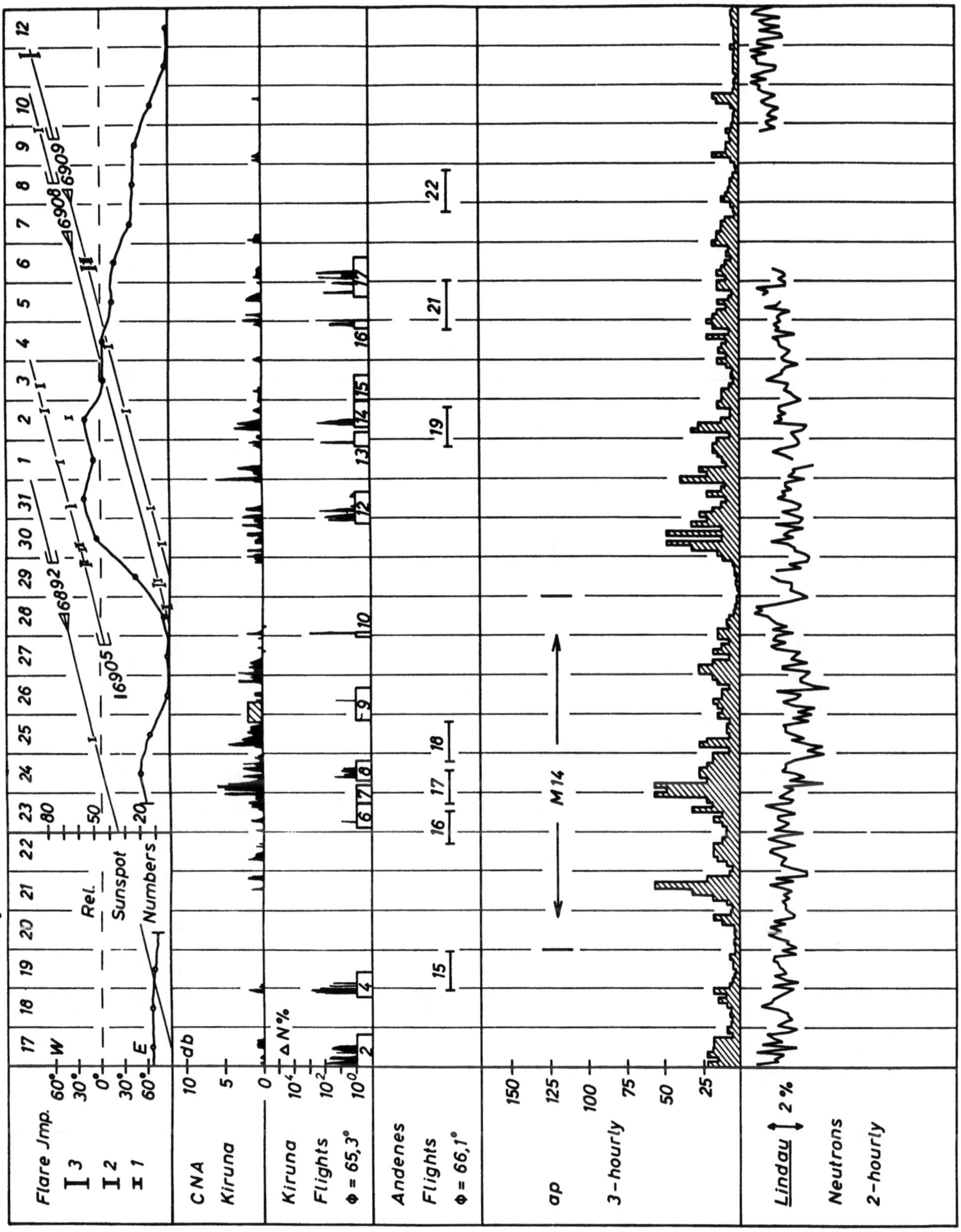

Fig. C2   Same as C1 for solar rotation 1779

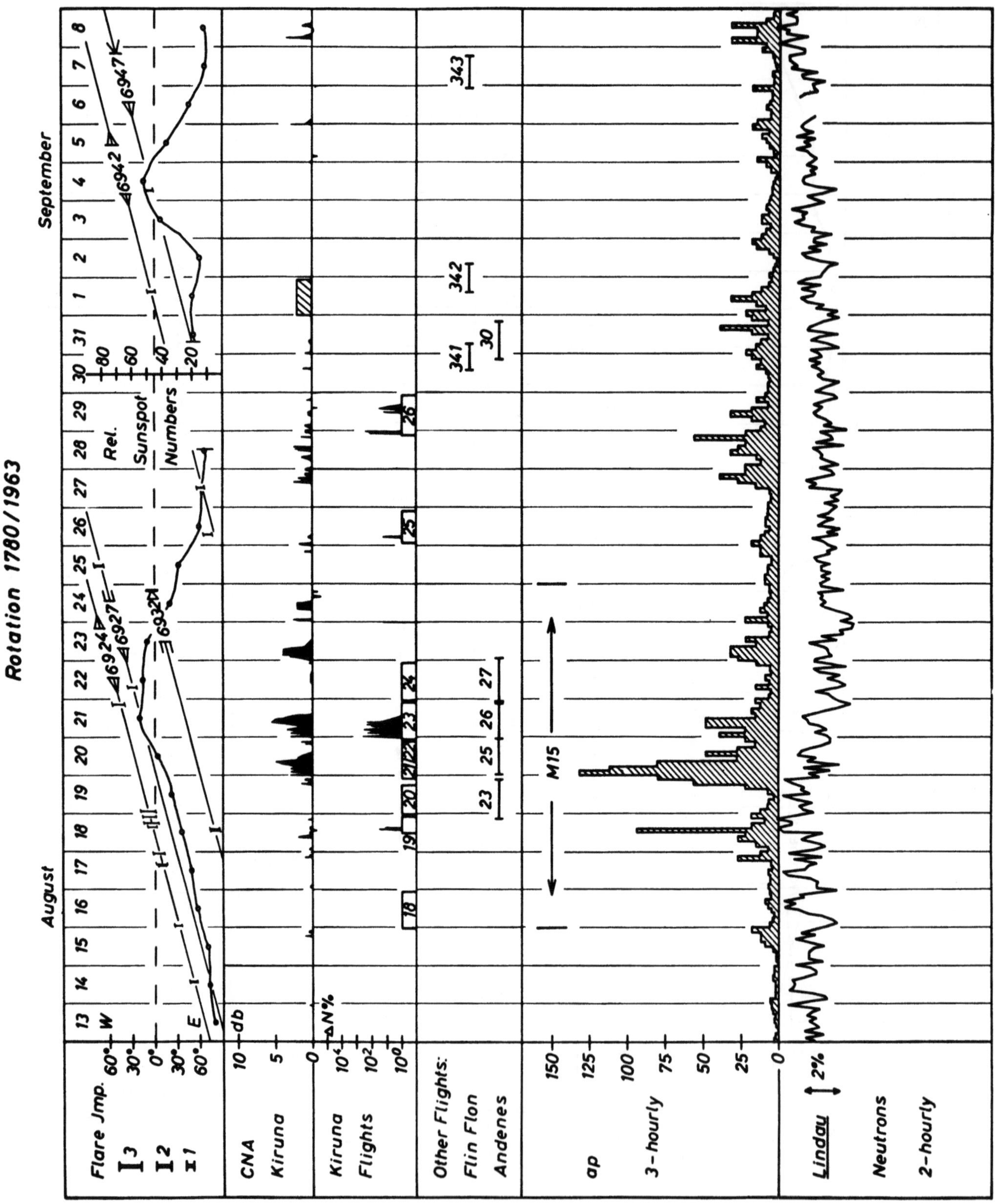

Fig. C3  Same as C1 for solar rotation 1780

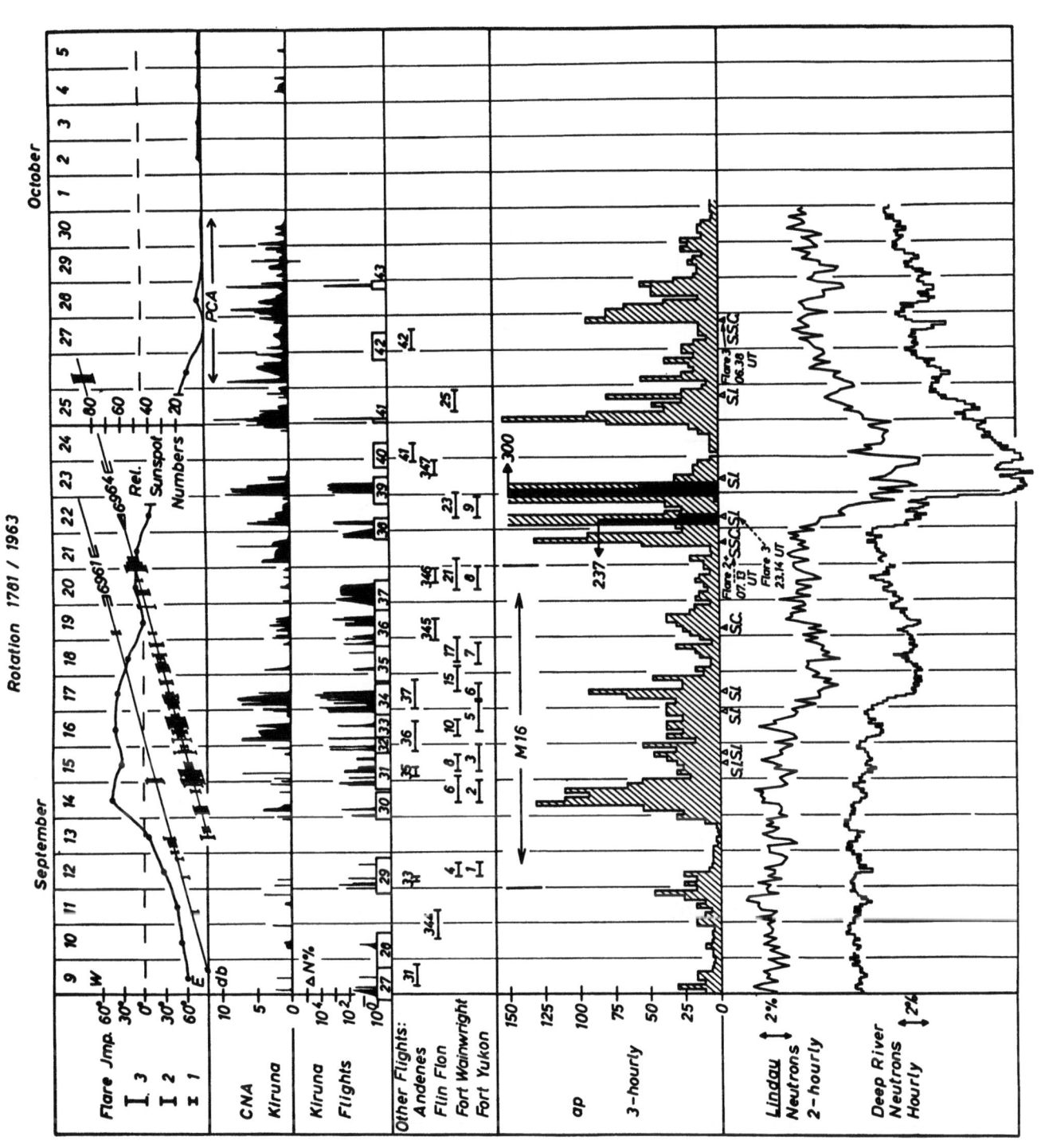

Fig. C4   Same as C1 for solar rotation 1781

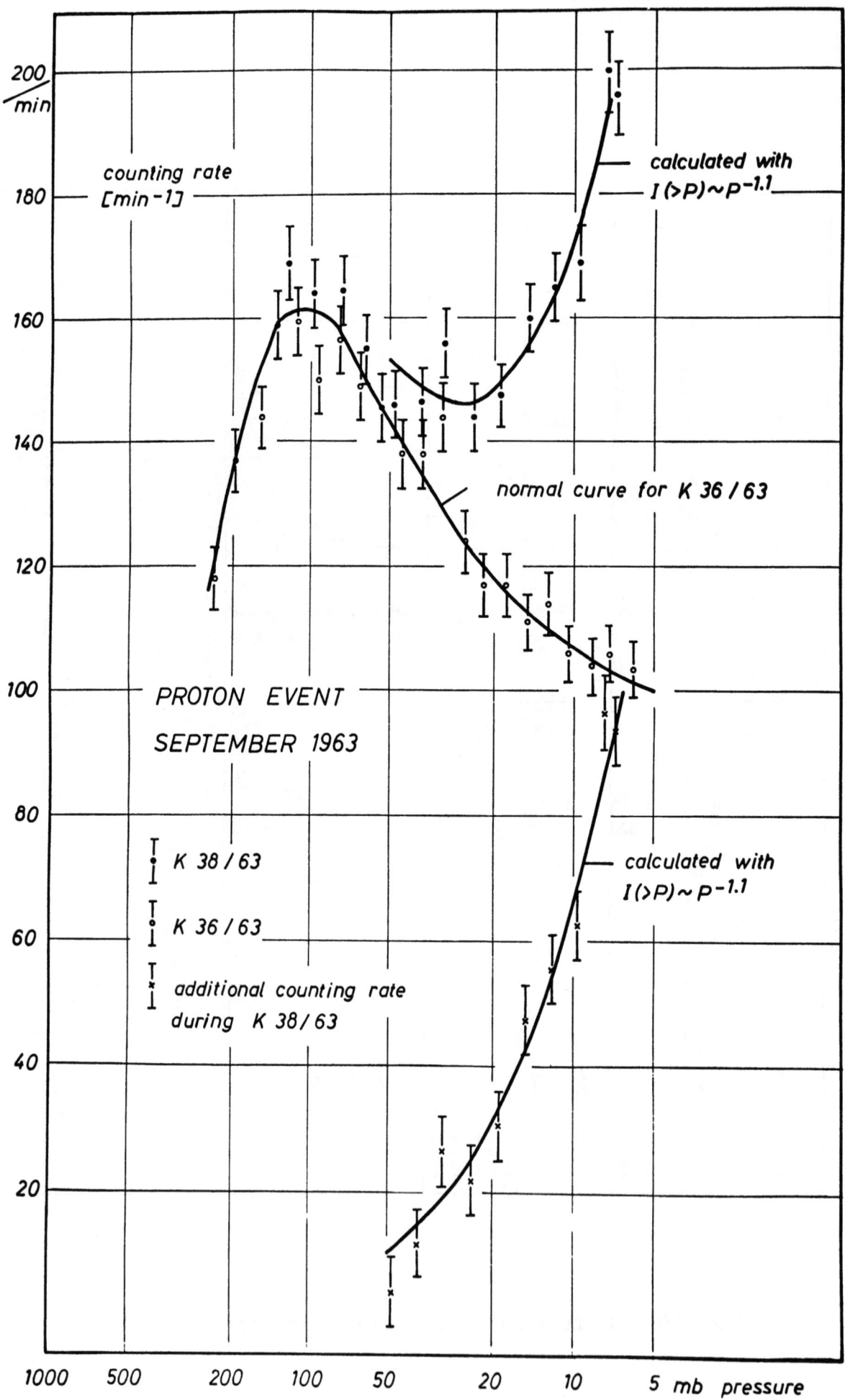

Fig. C5  Counting rates of the telescopes during the ascent of K36/63 and K38/63 and the additional counting rate of the telescope during K38/63
($I(>P)$ = integral flux, $P$ = pressure)

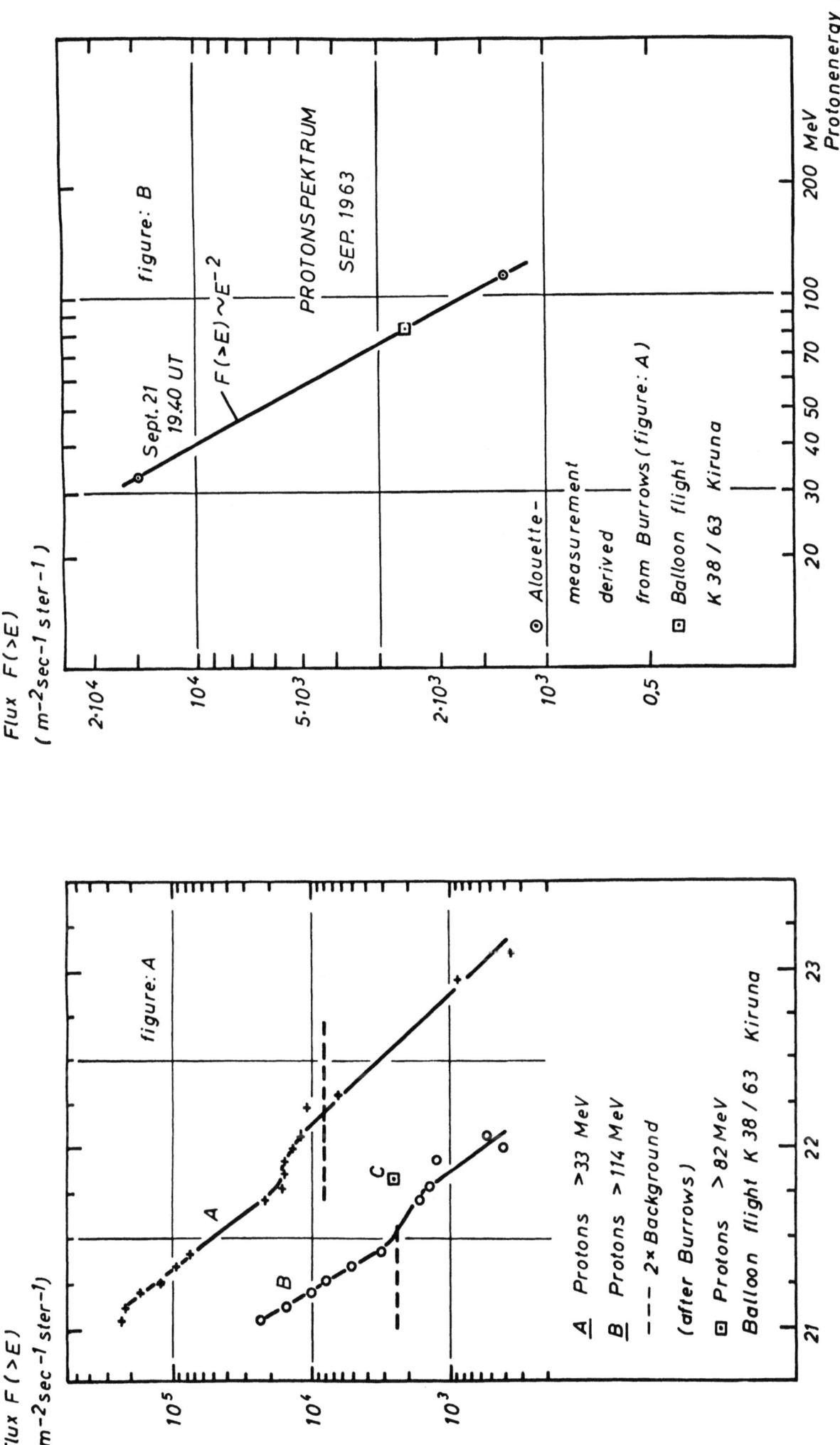

Fig. C6  A : Proton fluxes measured by the Alouette satellite (Burrows 1964) and by the balloon K38/63
B : Proton spectrum at 1940 UT on September 21, 1963

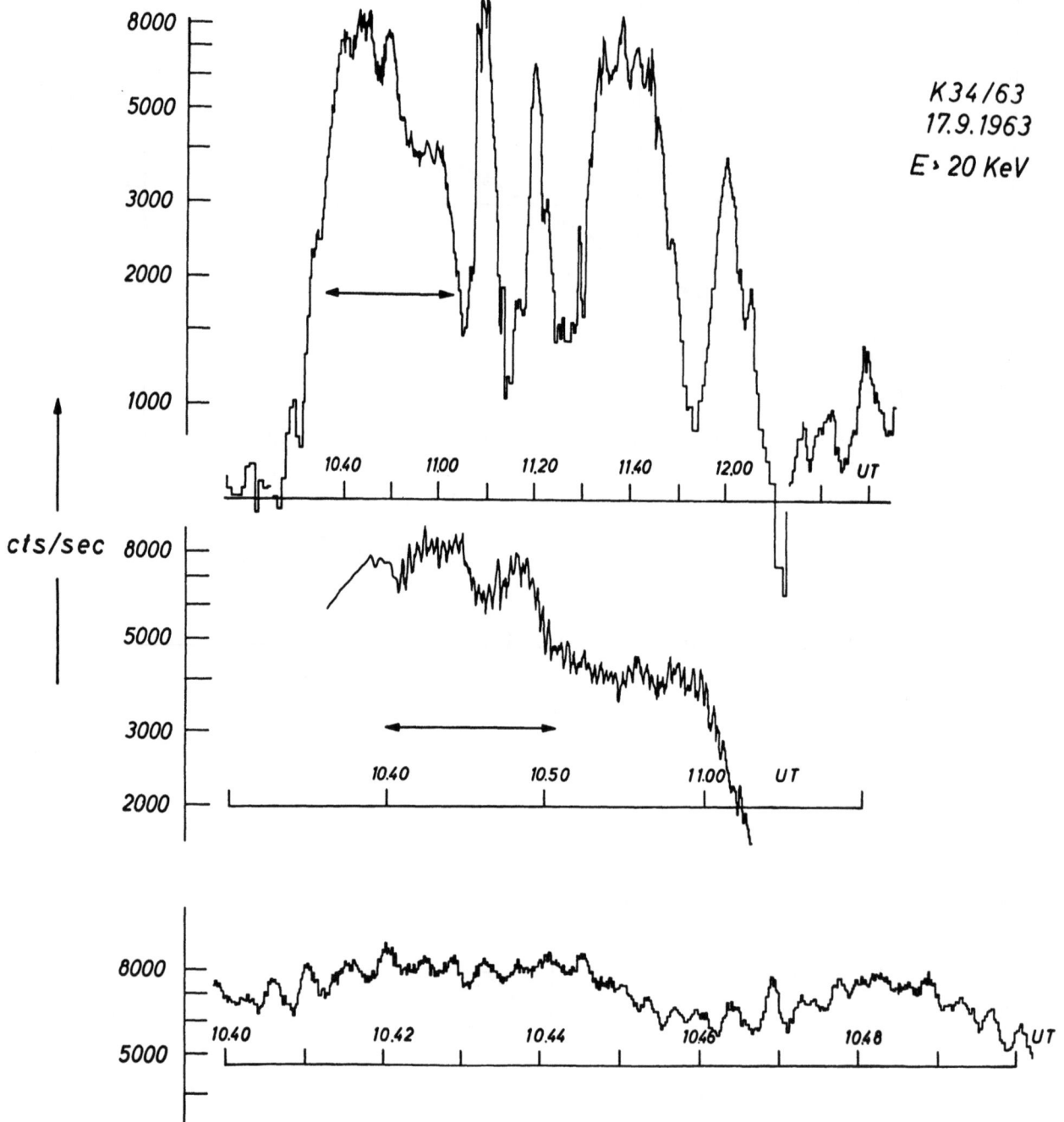

Fig. C7  Example of quasiperiodic intensity variations of the x-ray flux with an apparent period of 24 sec on different time scales. The arrows show the periods represented in the next lower diagram respectively.

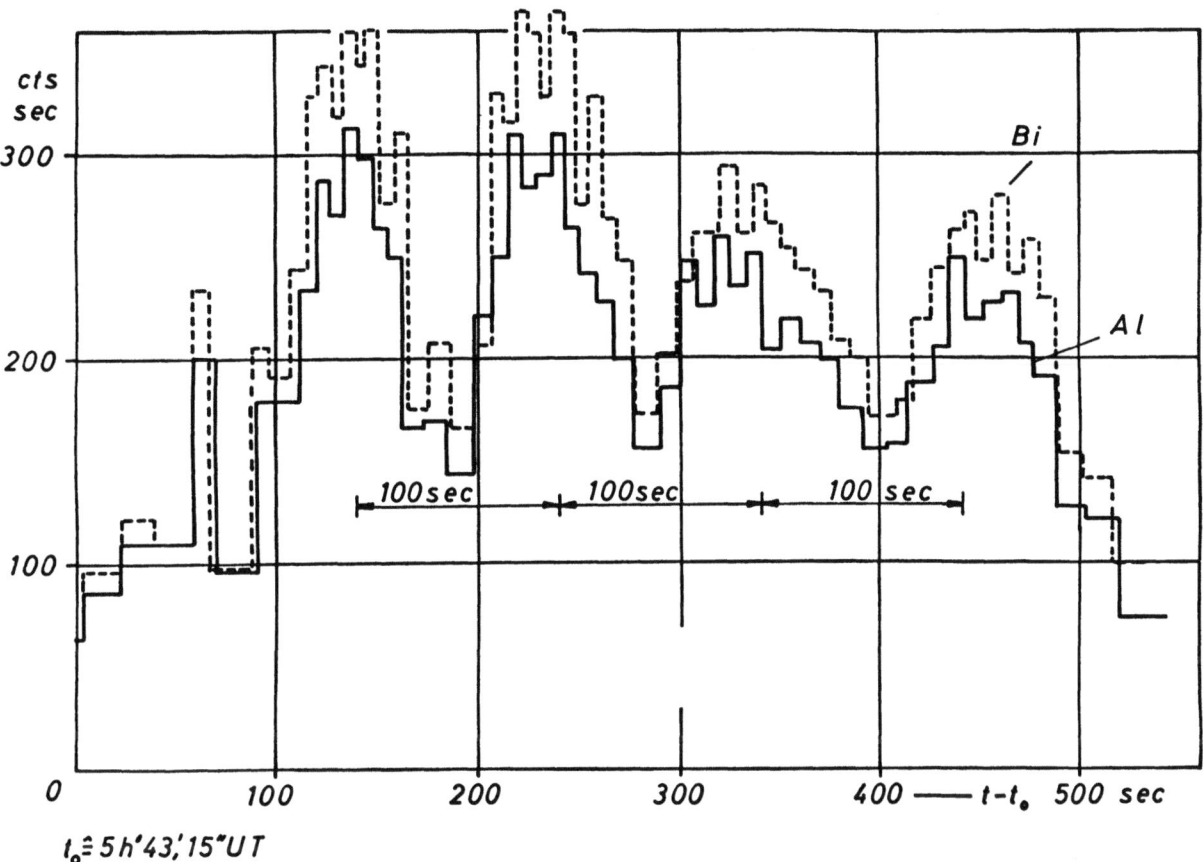

Fig. C8  Example of quasiperiodic intensity variations of the x-ray flux with an apparent period of 100 sec.

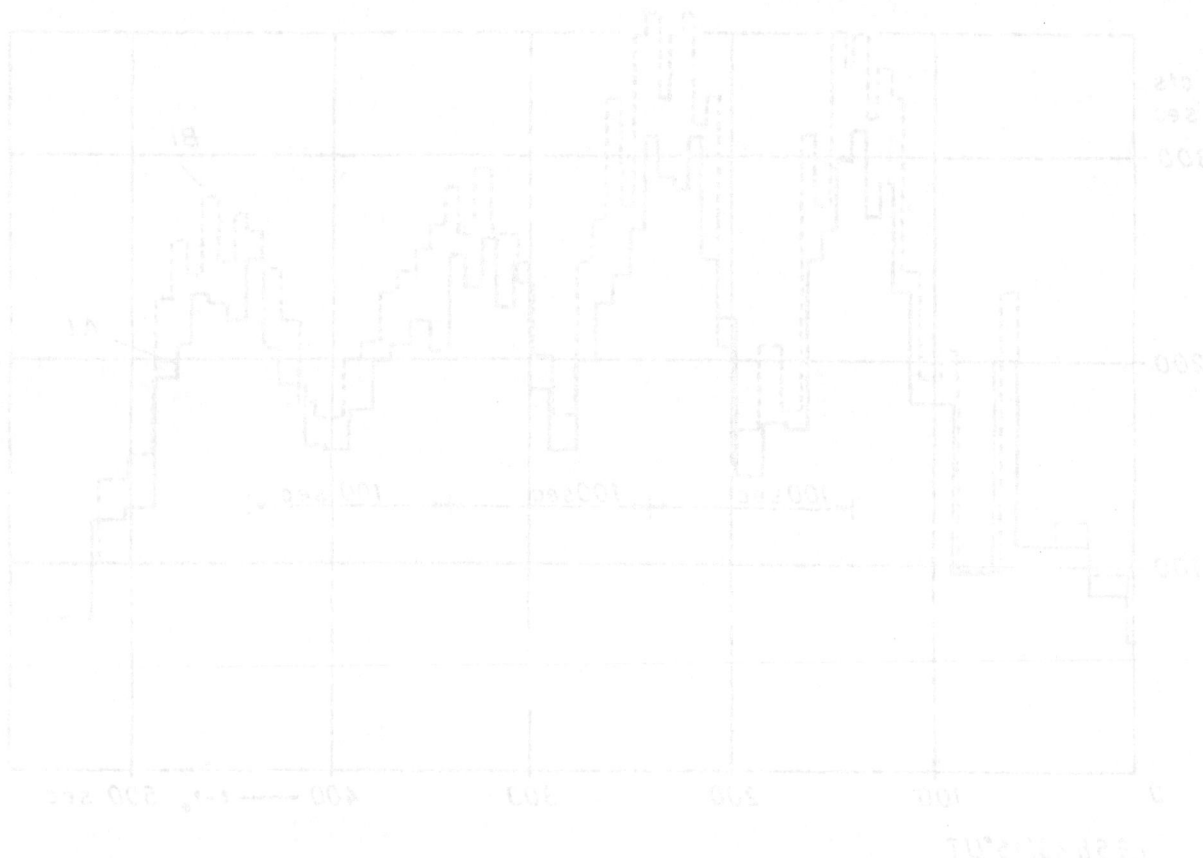

Fig. 68 Example of quasiperiodic anomaly variations of the X-ray flux with an apparent period of 100 sec.

Balloon Flights 1962

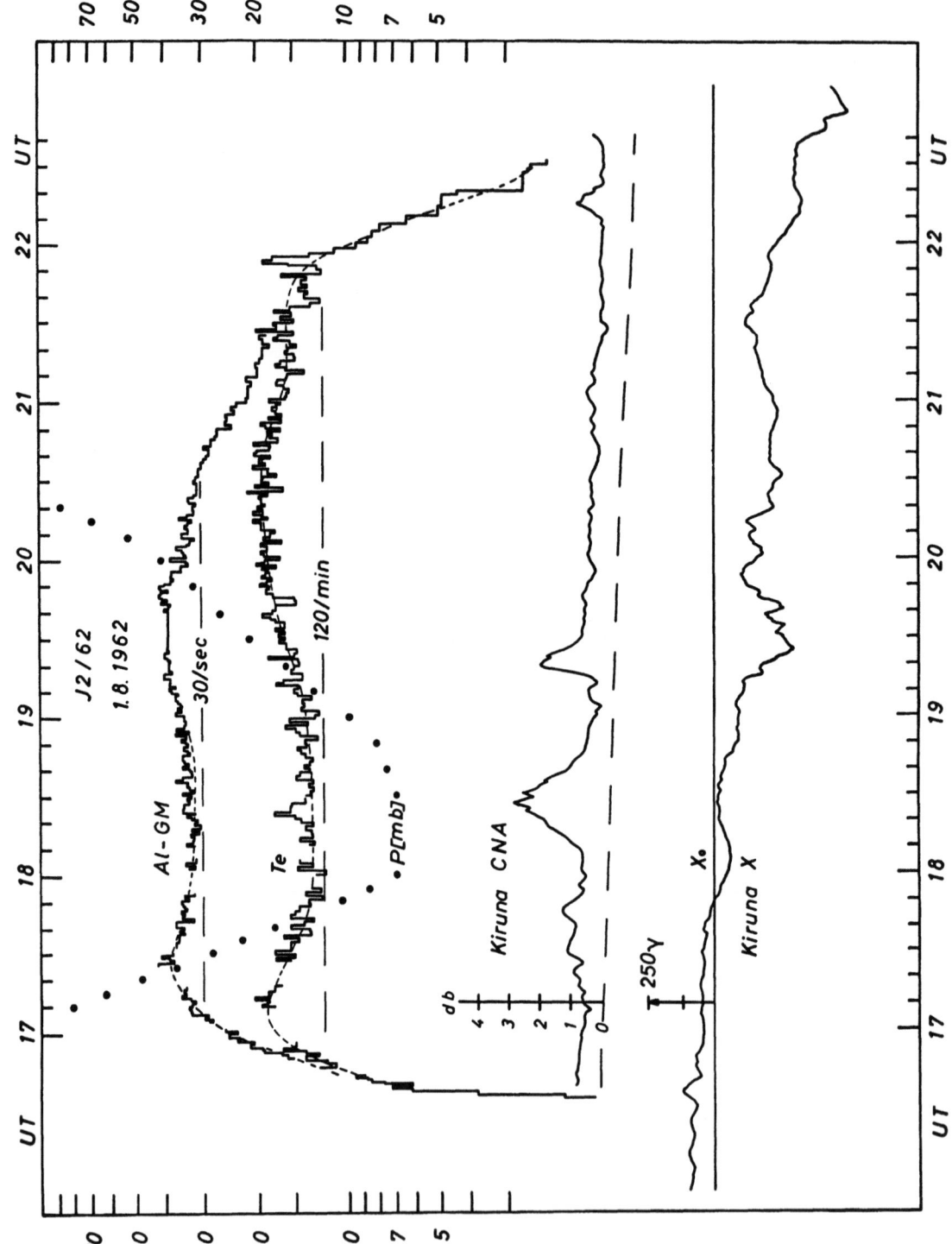

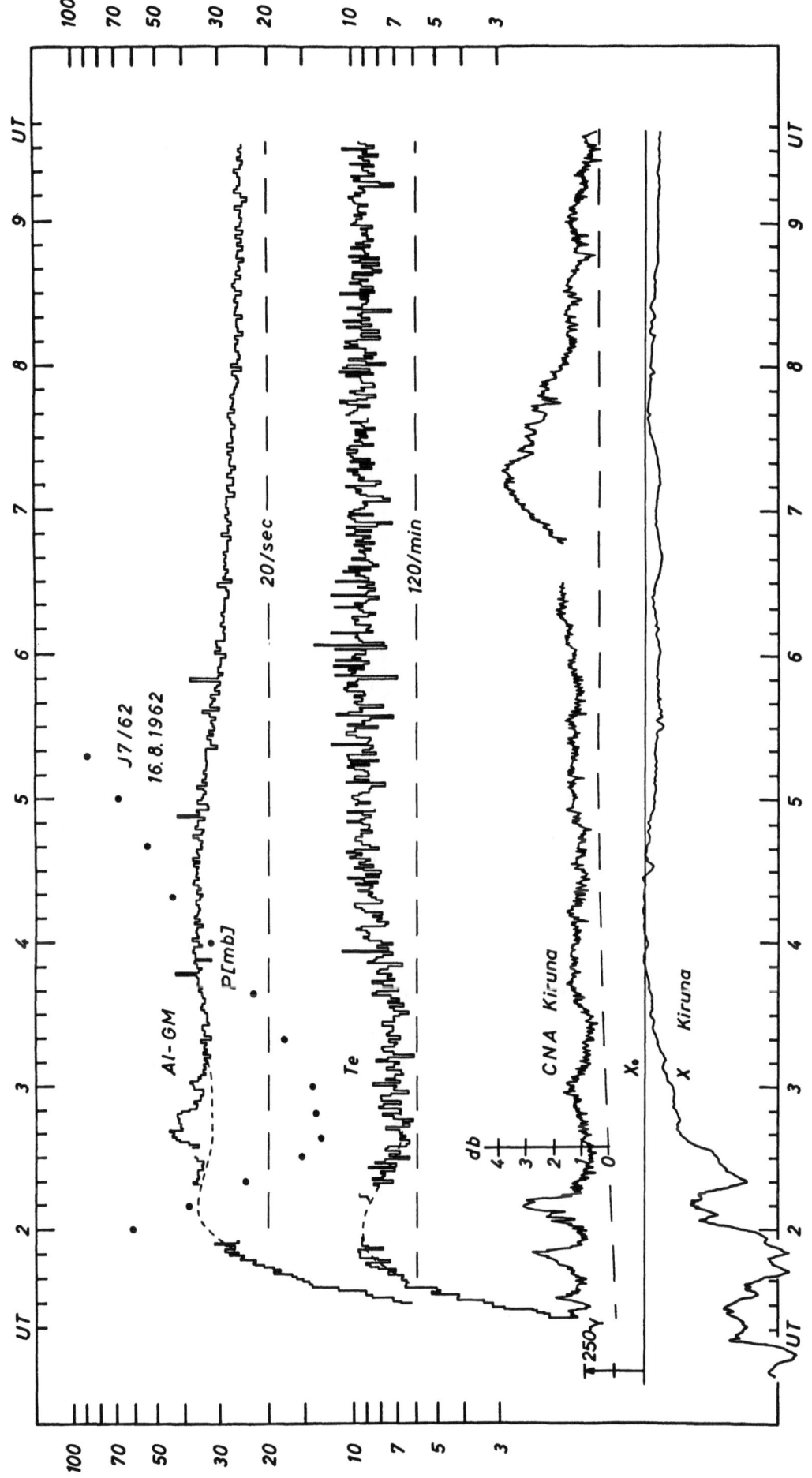

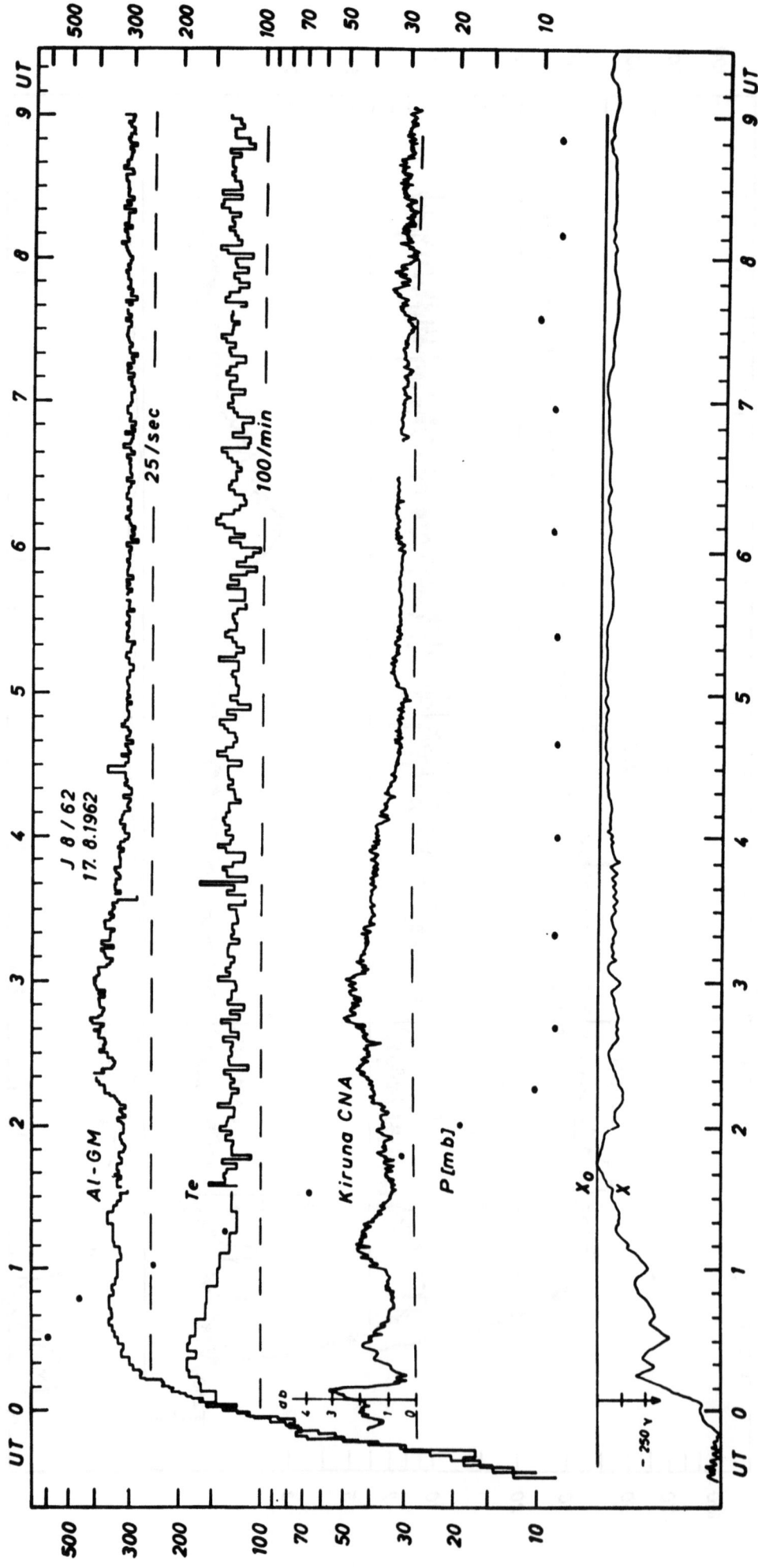

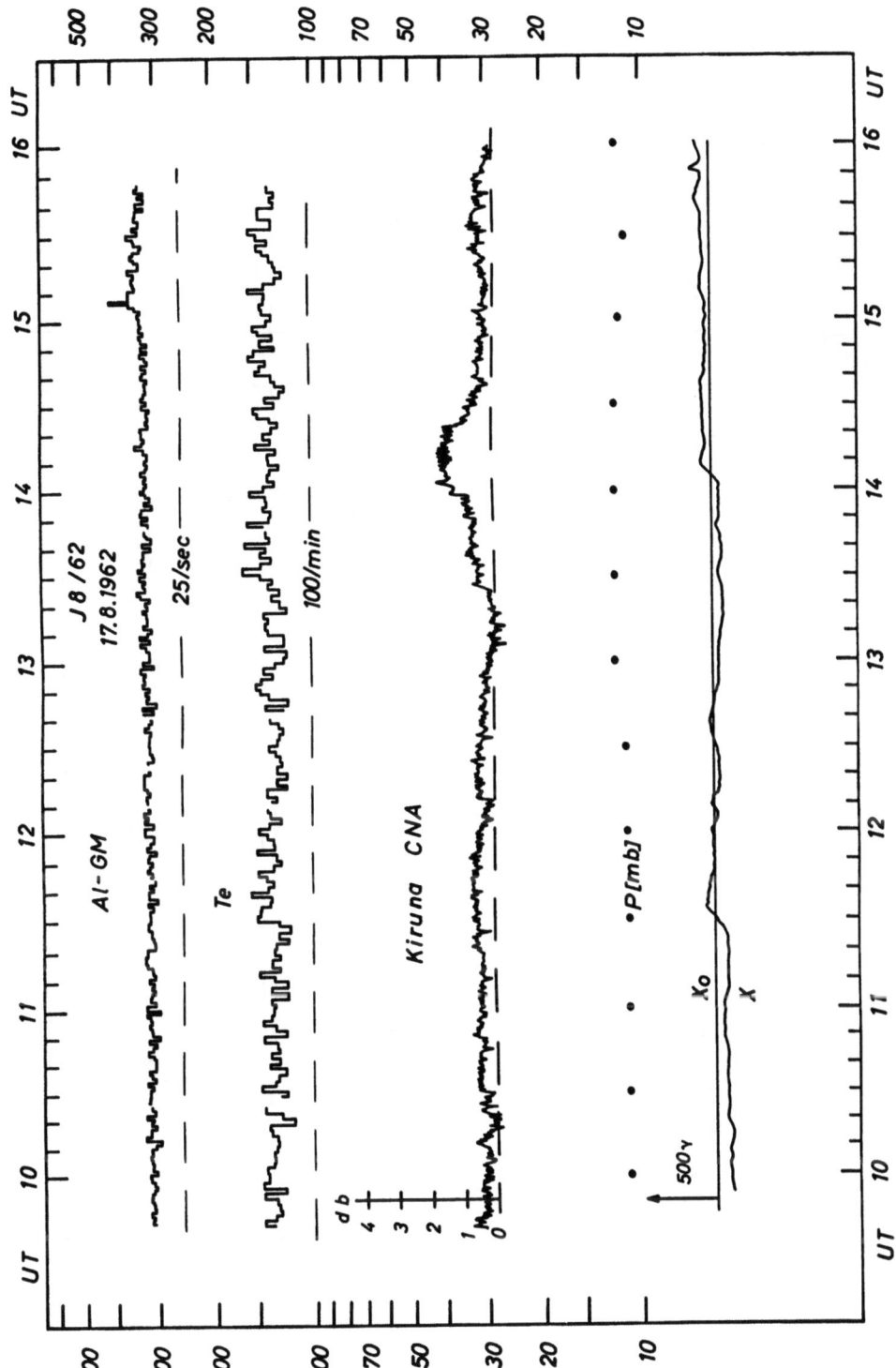

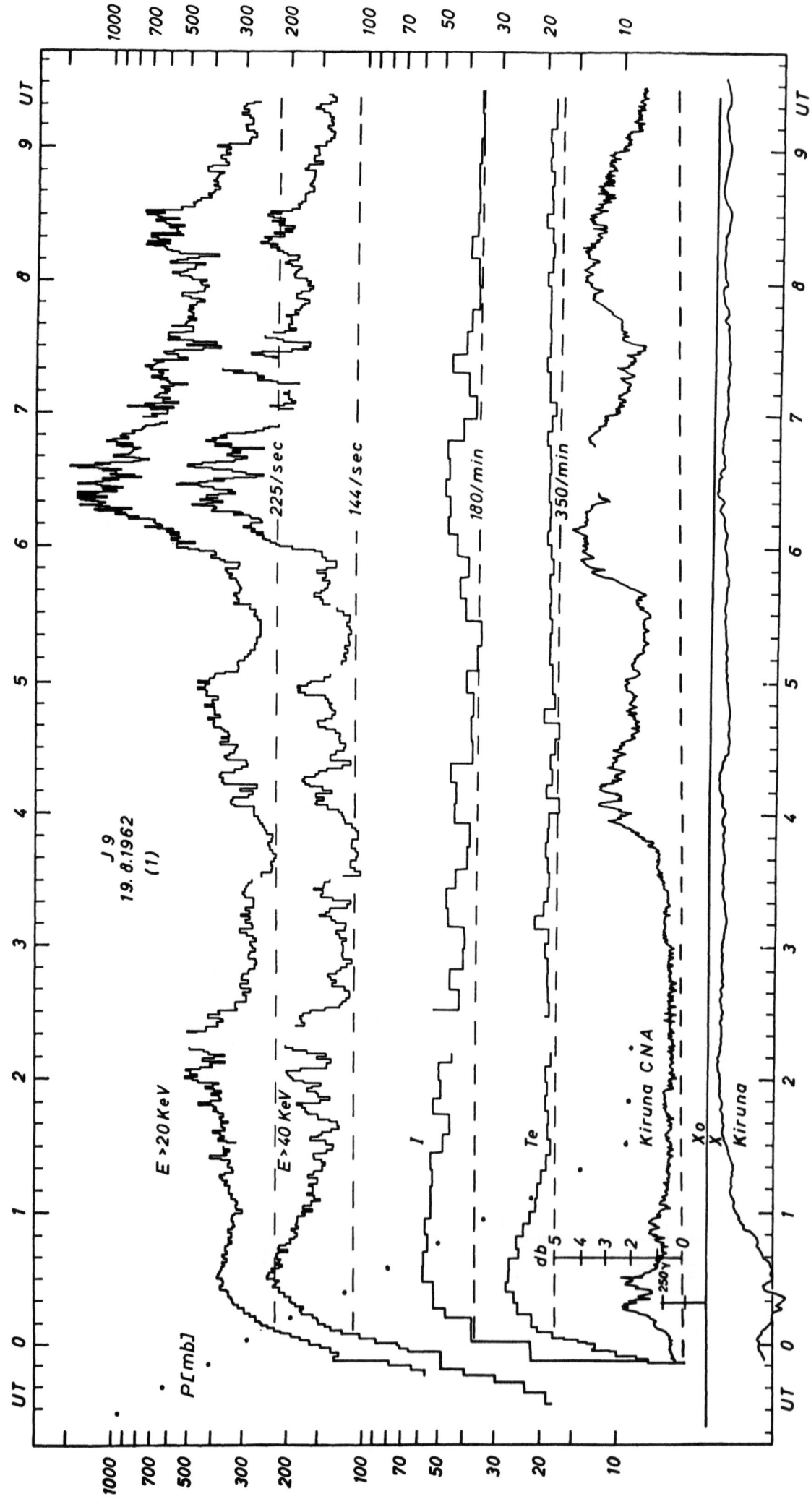

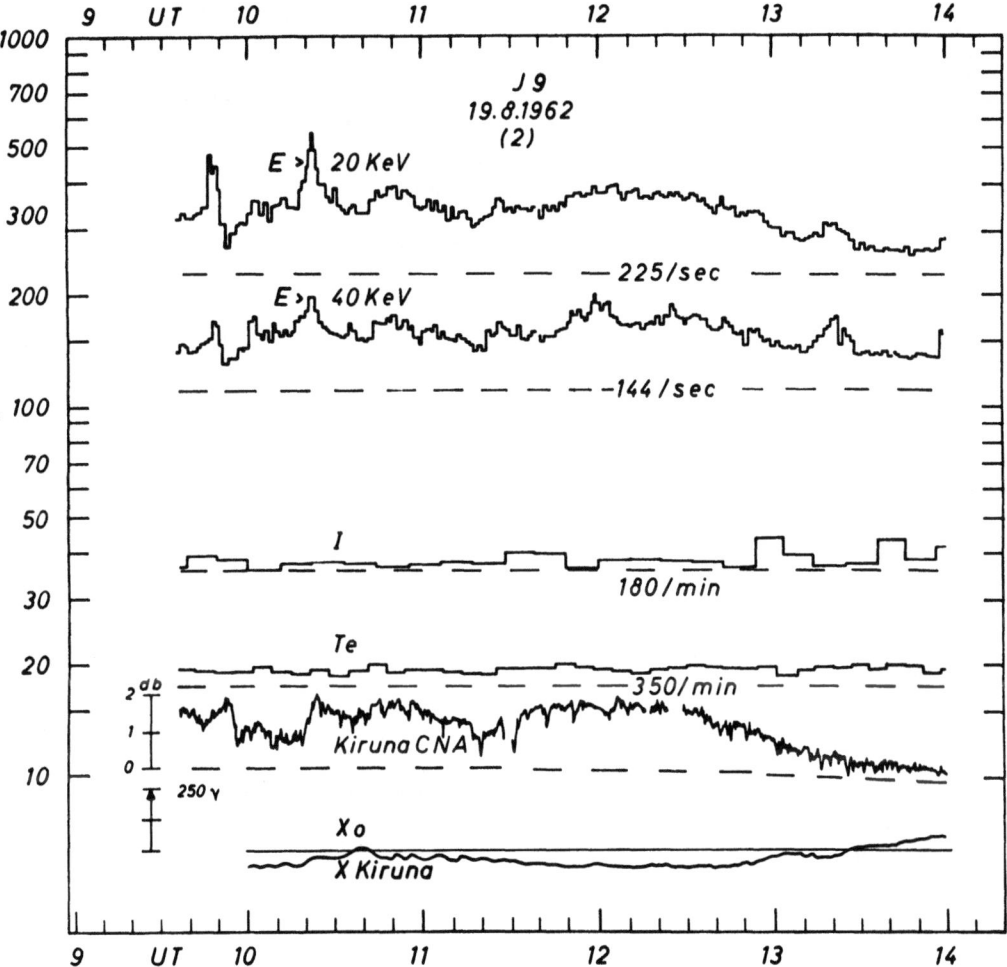

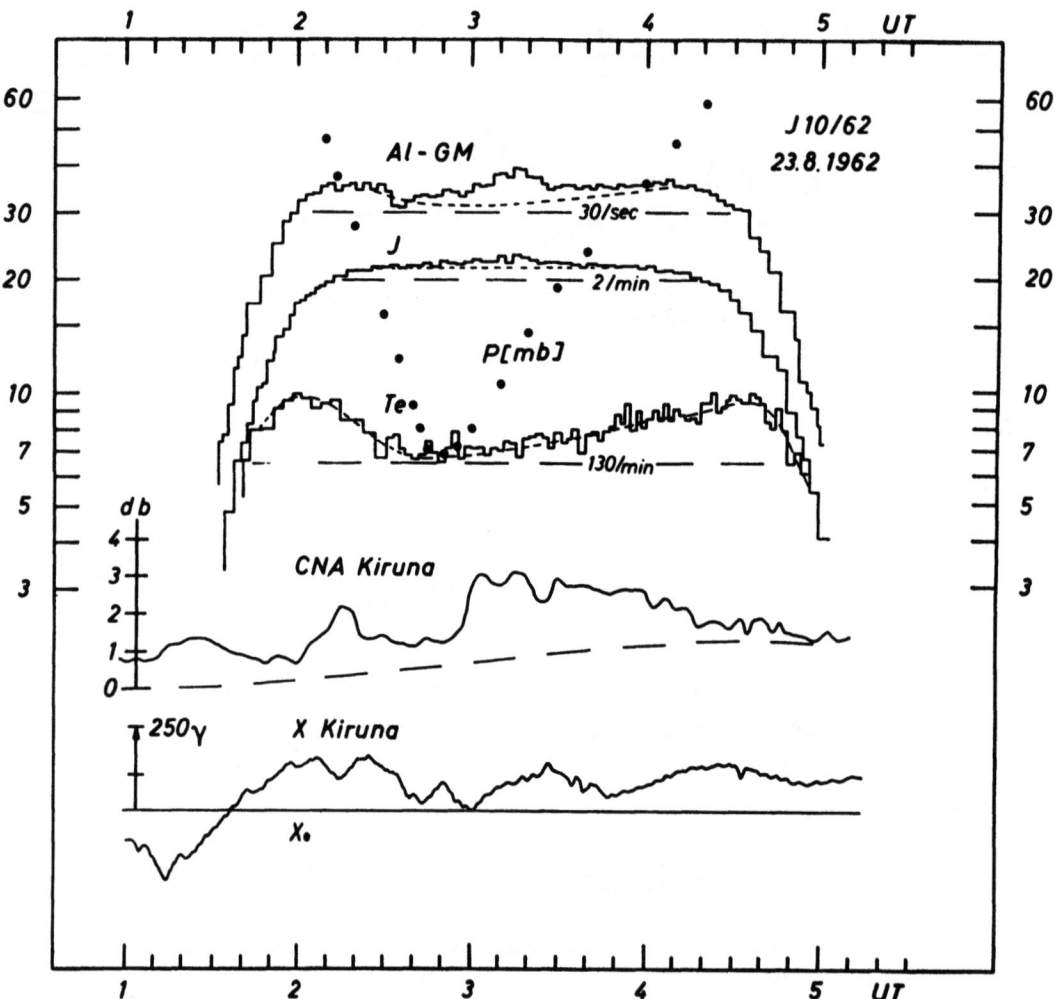

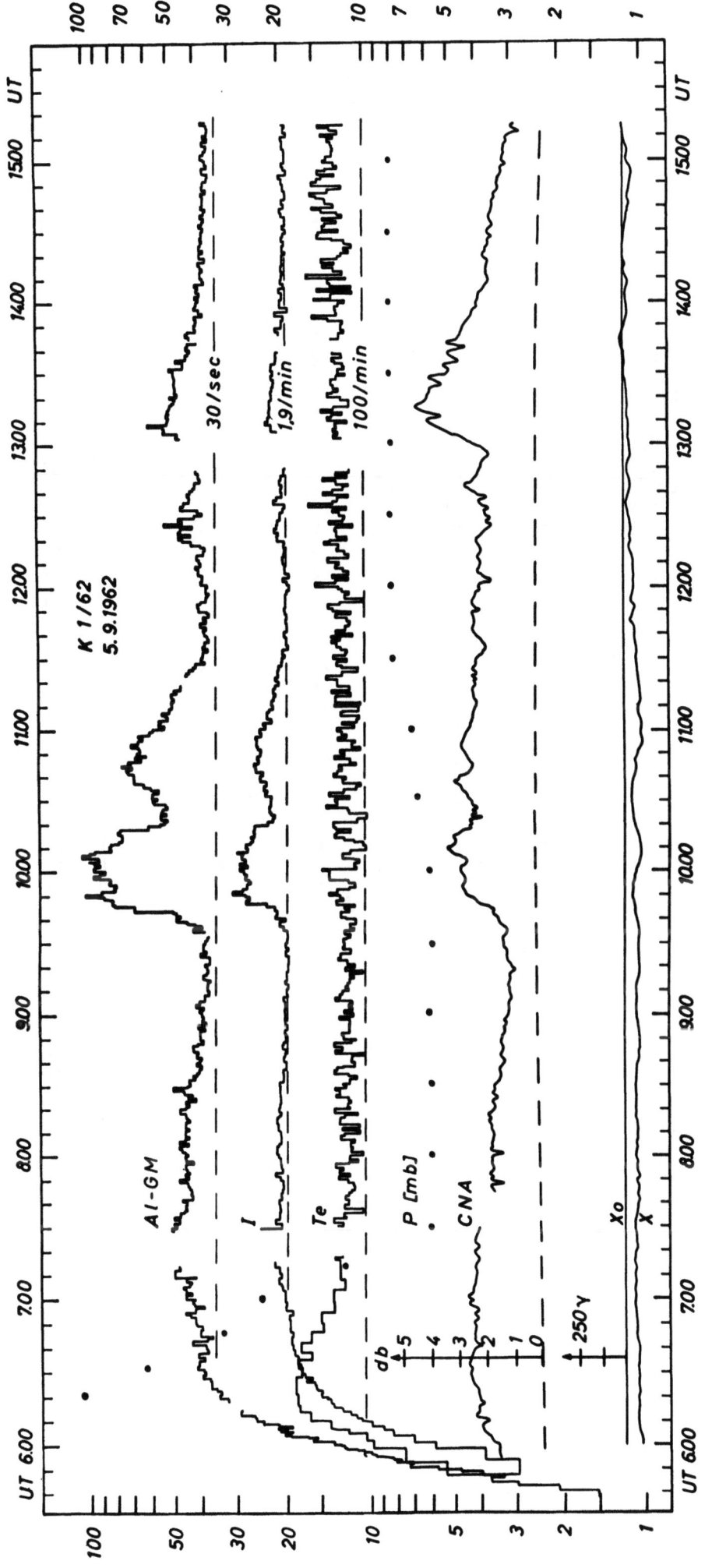

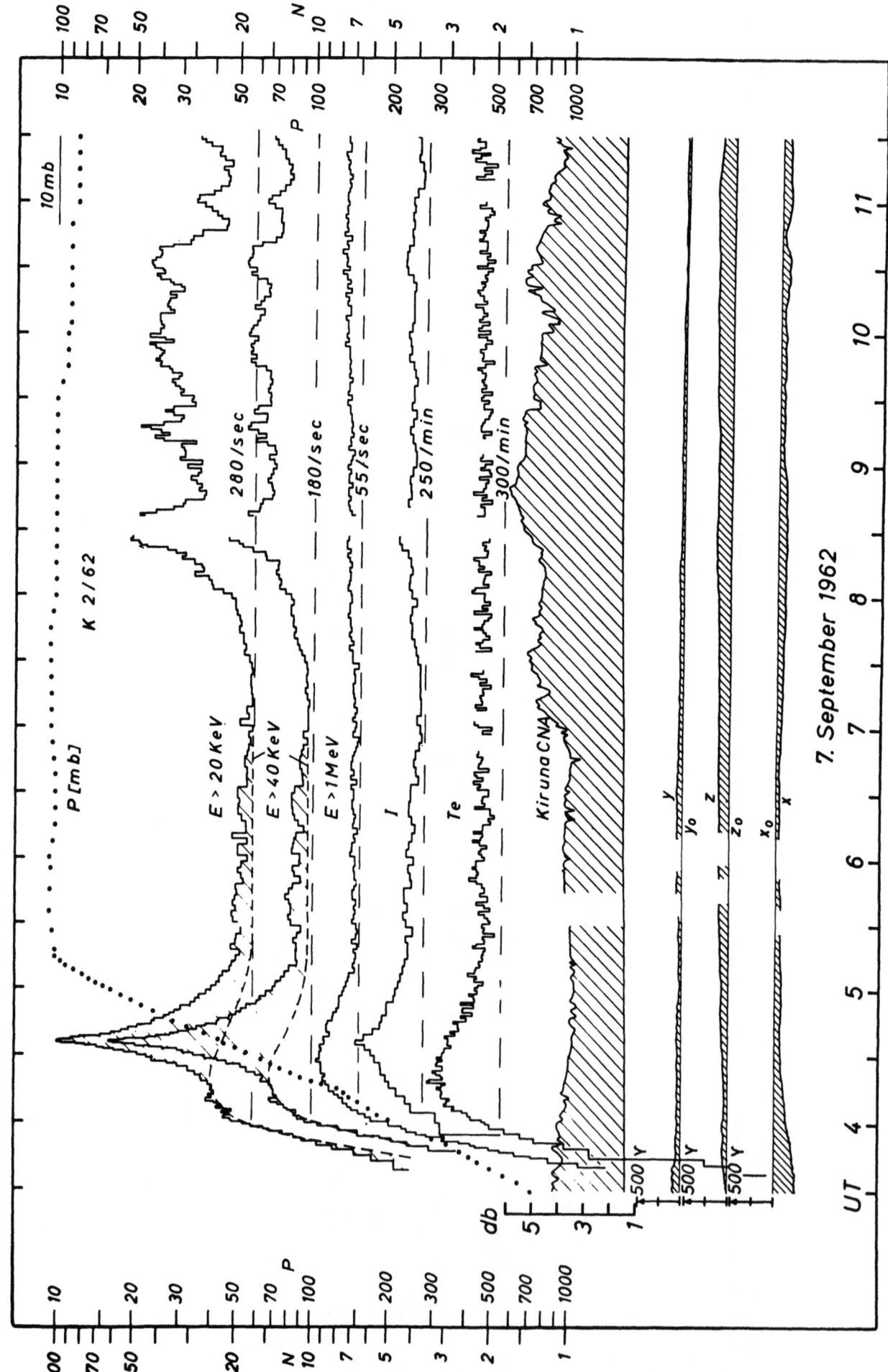

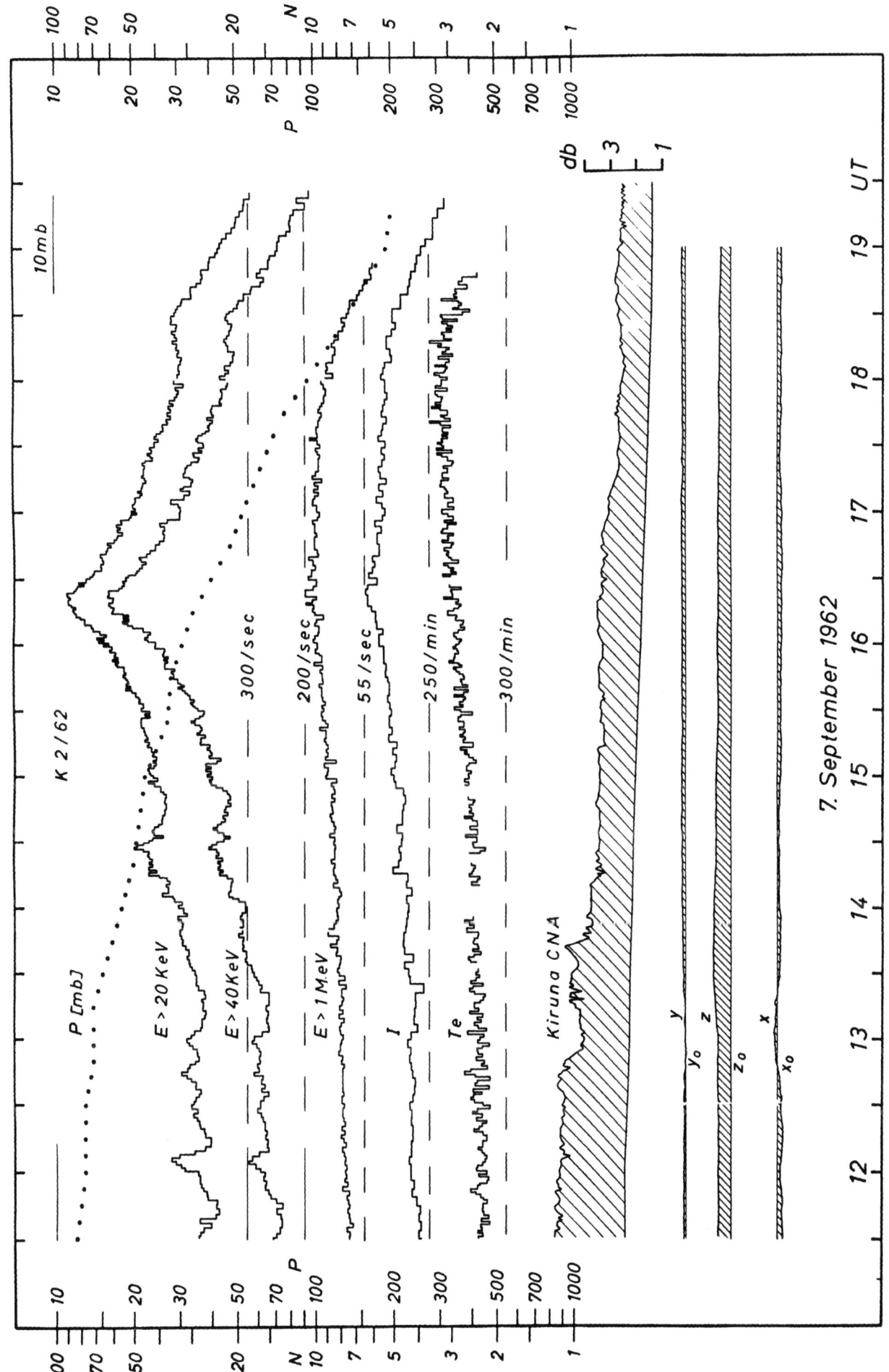

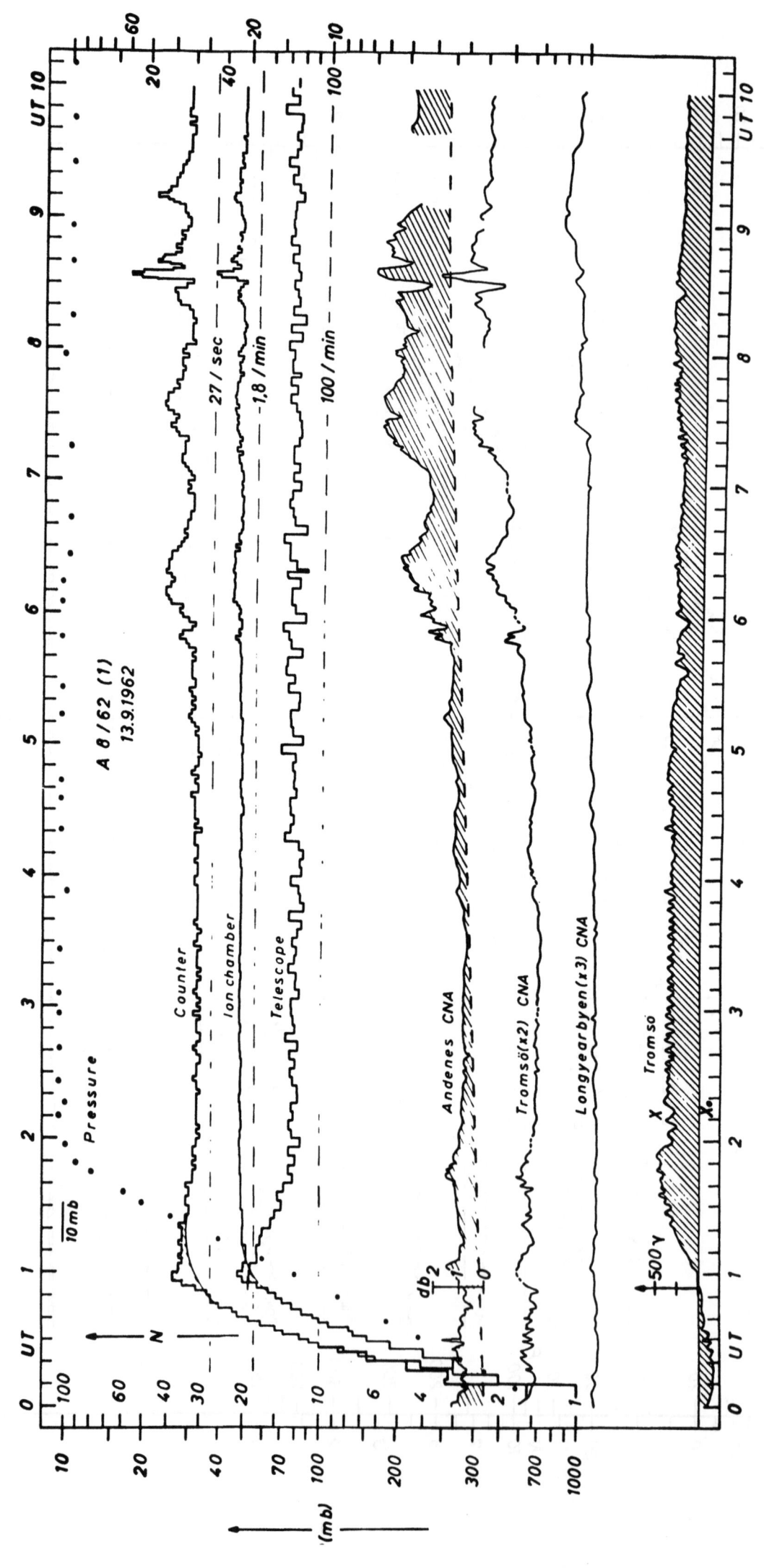

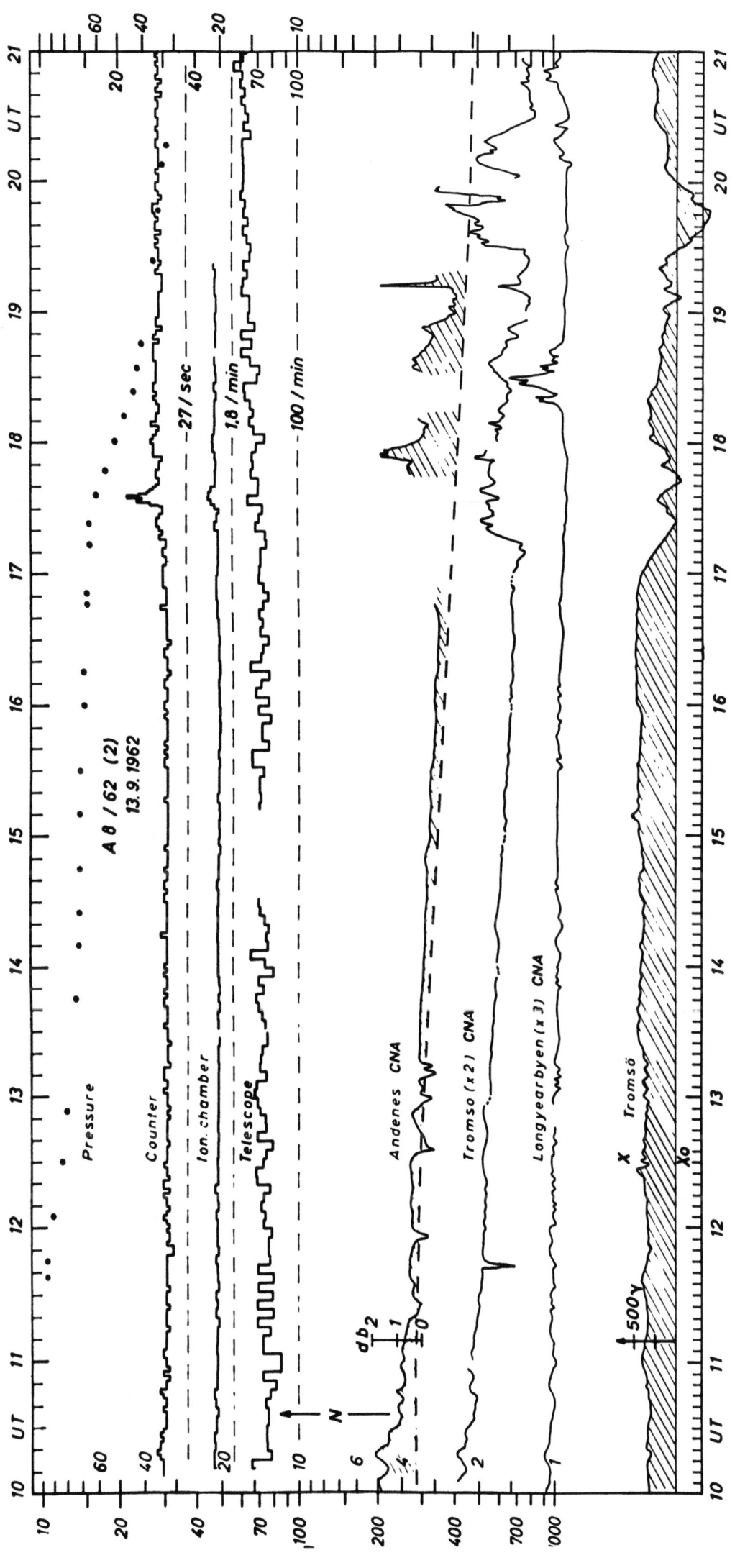

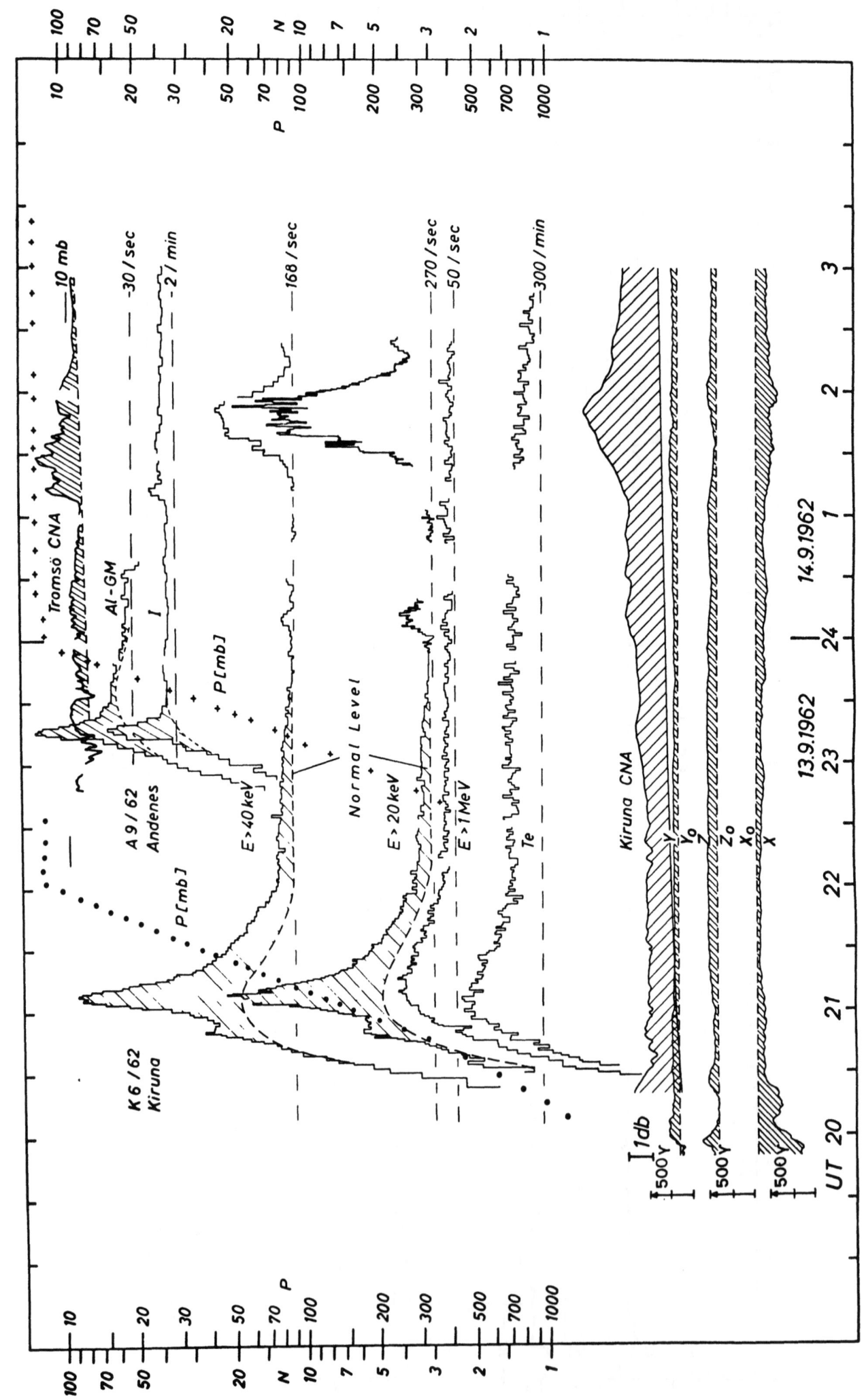

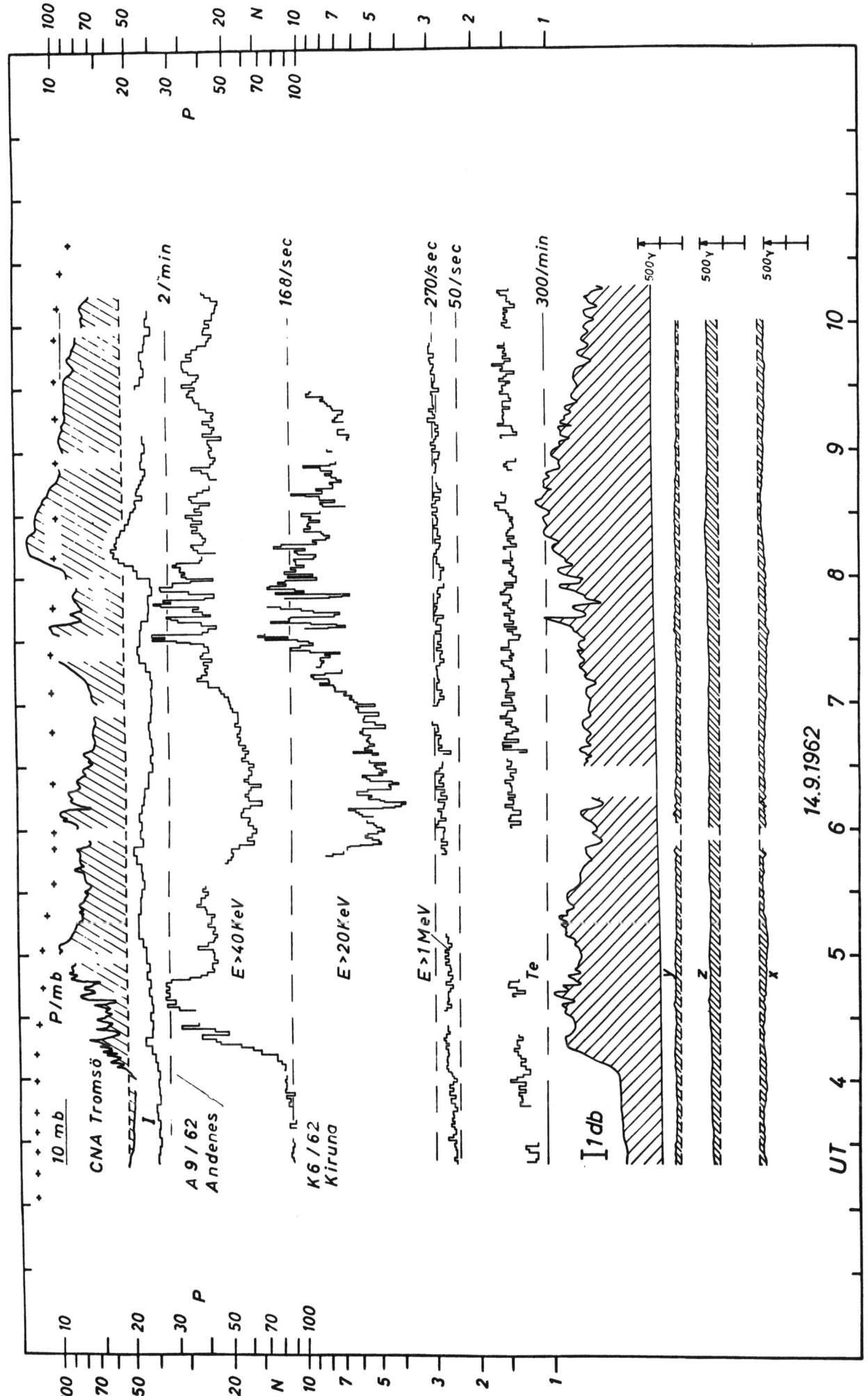

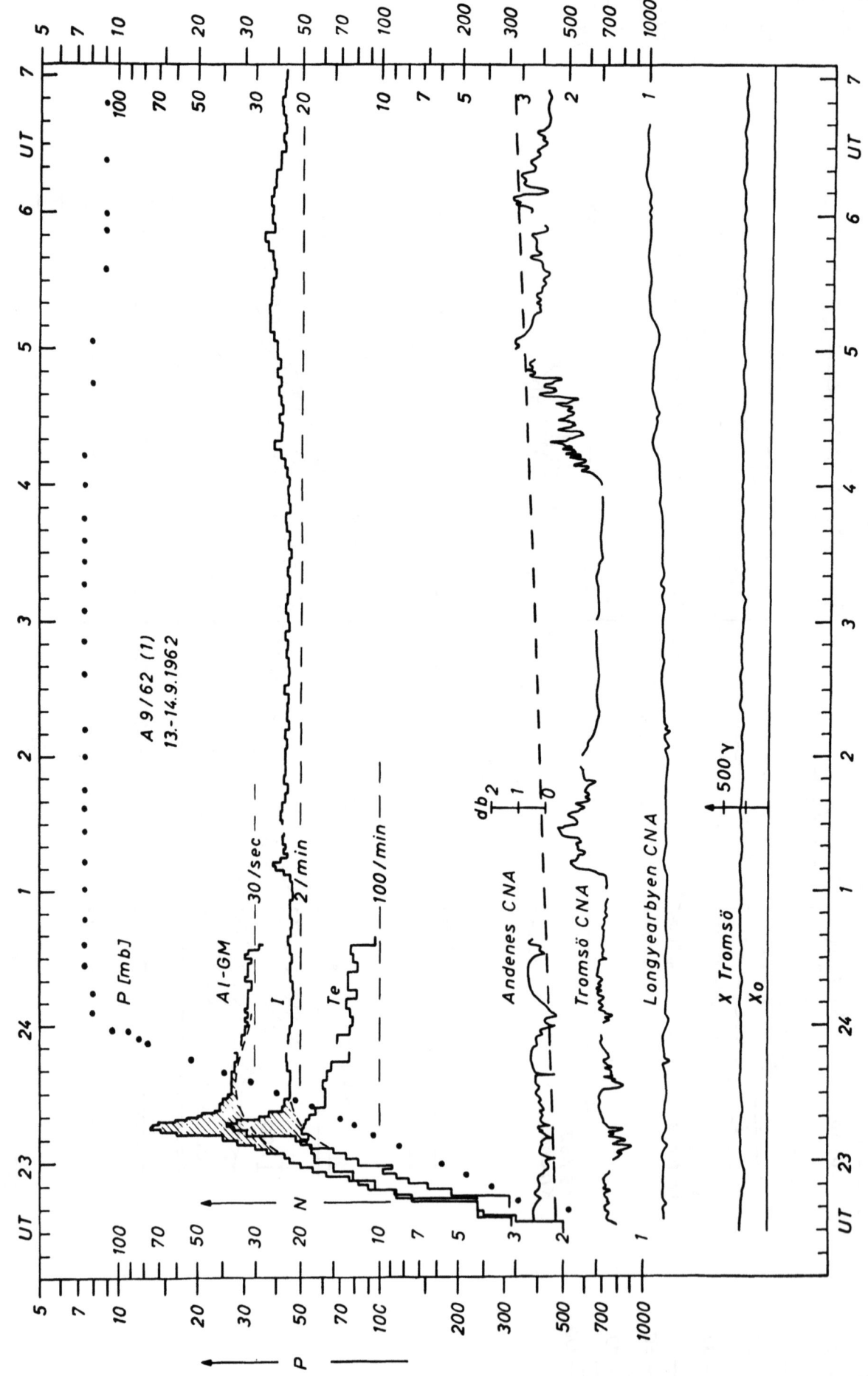

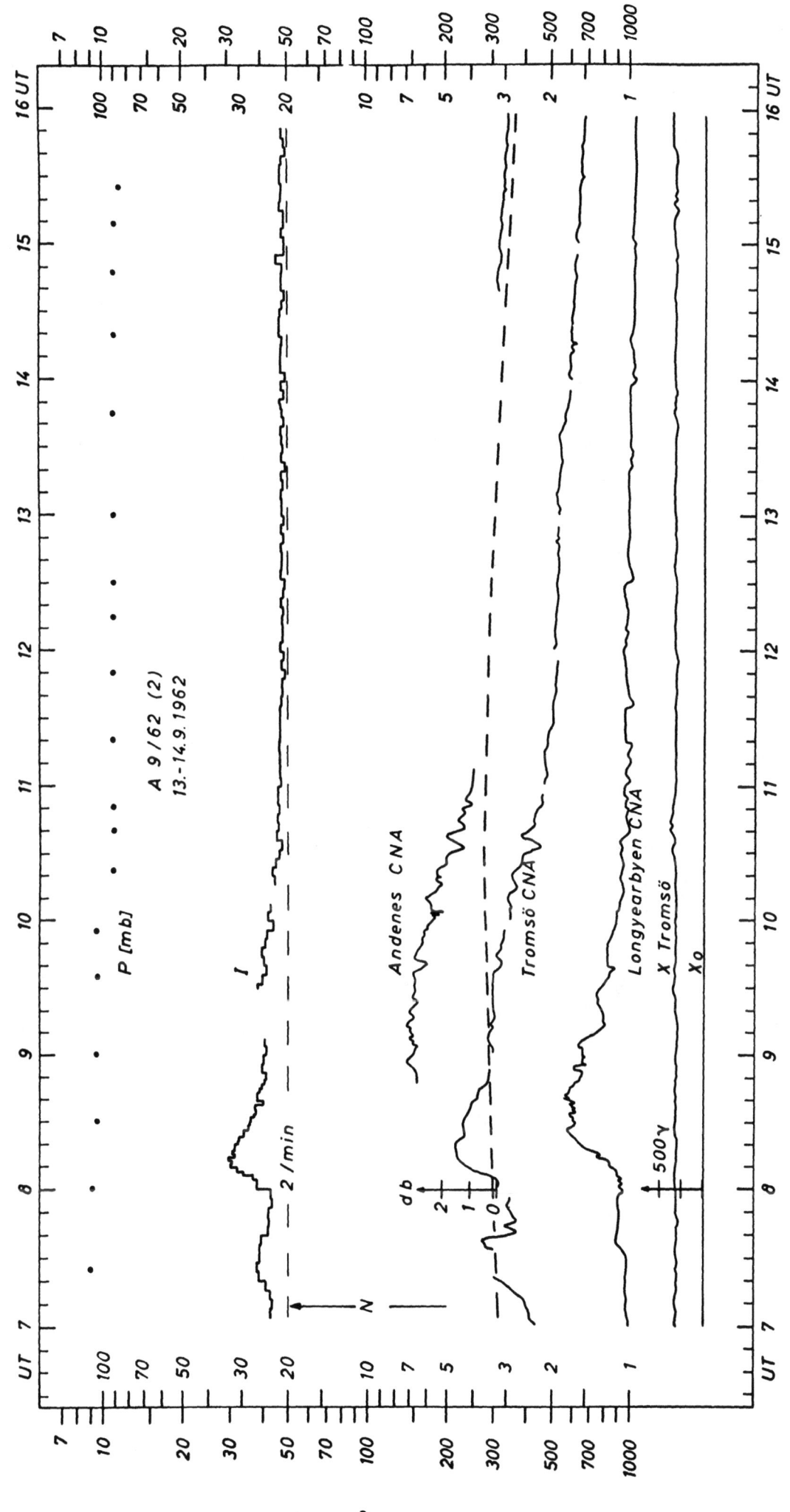

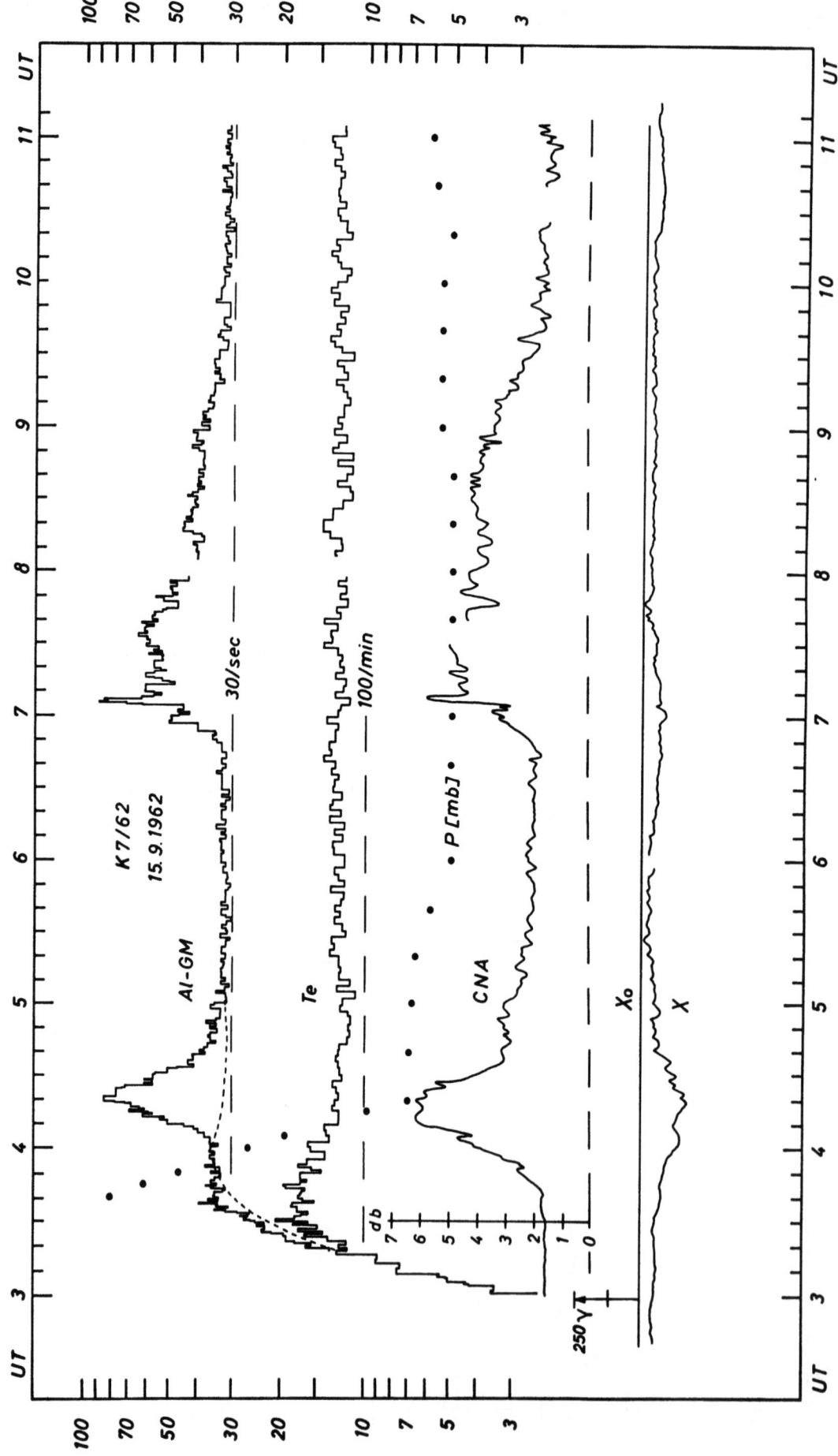

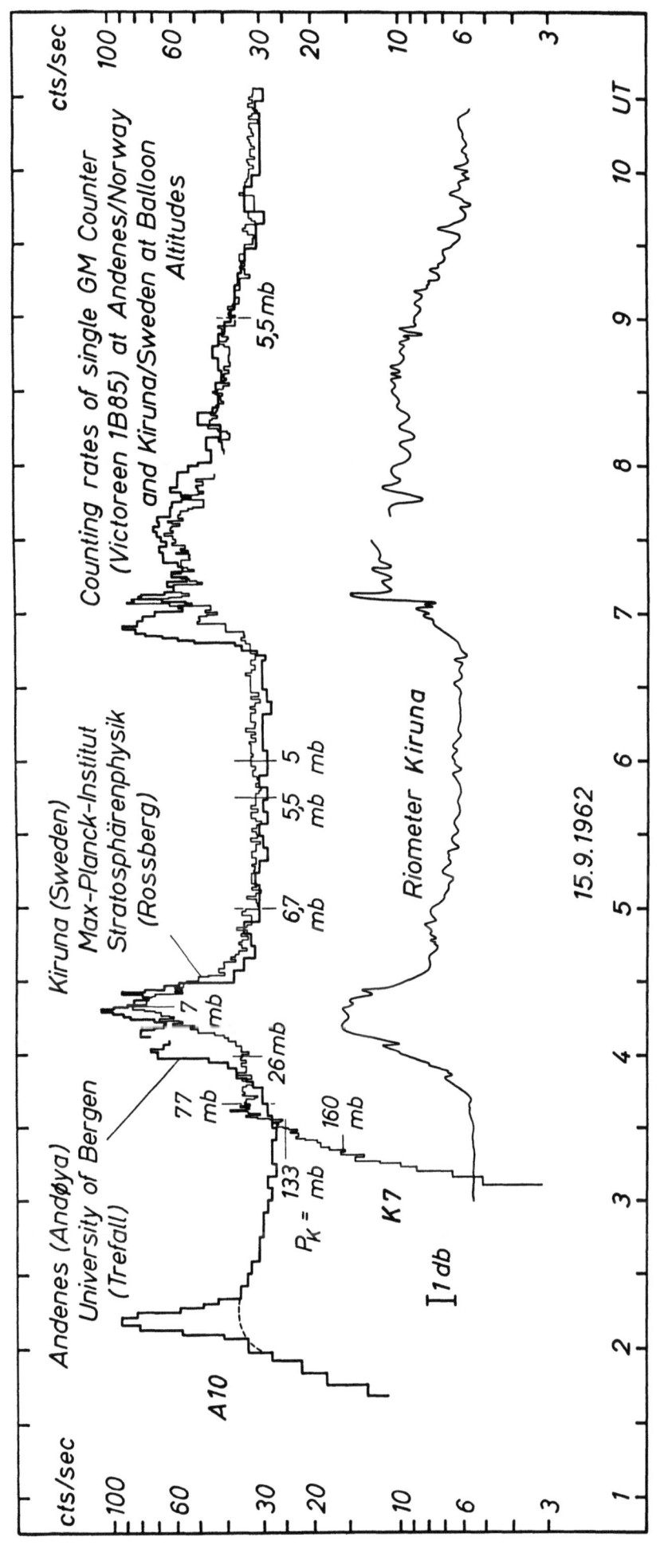

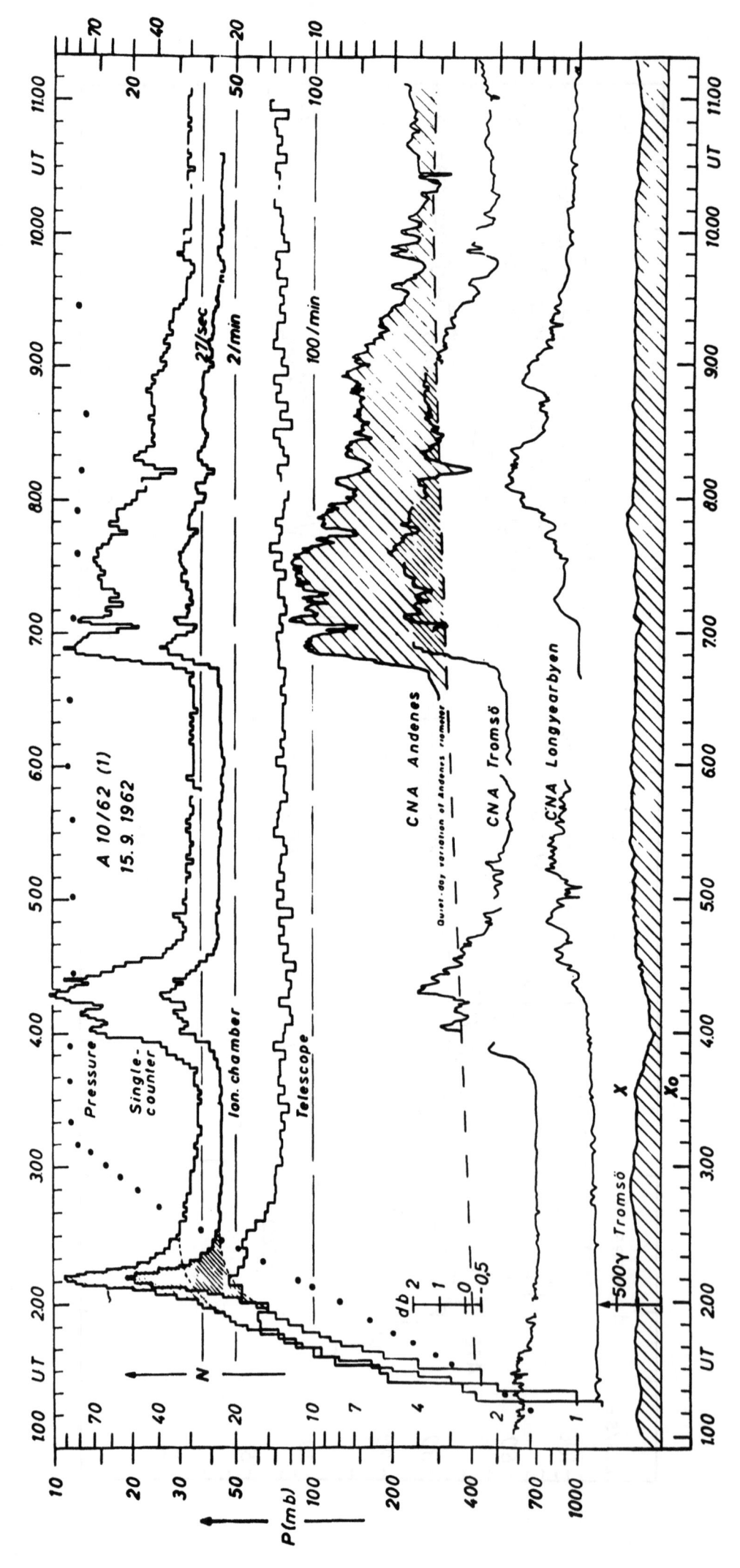

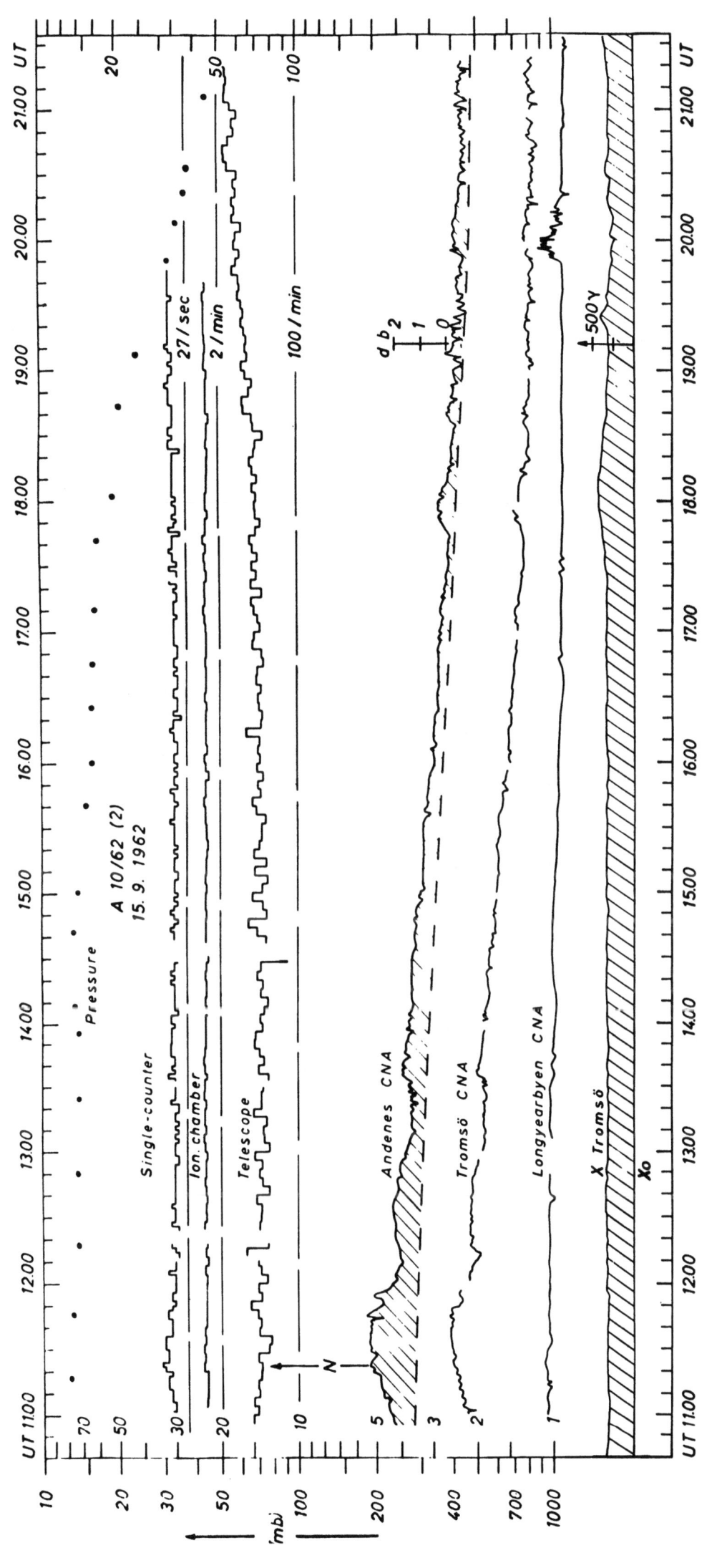

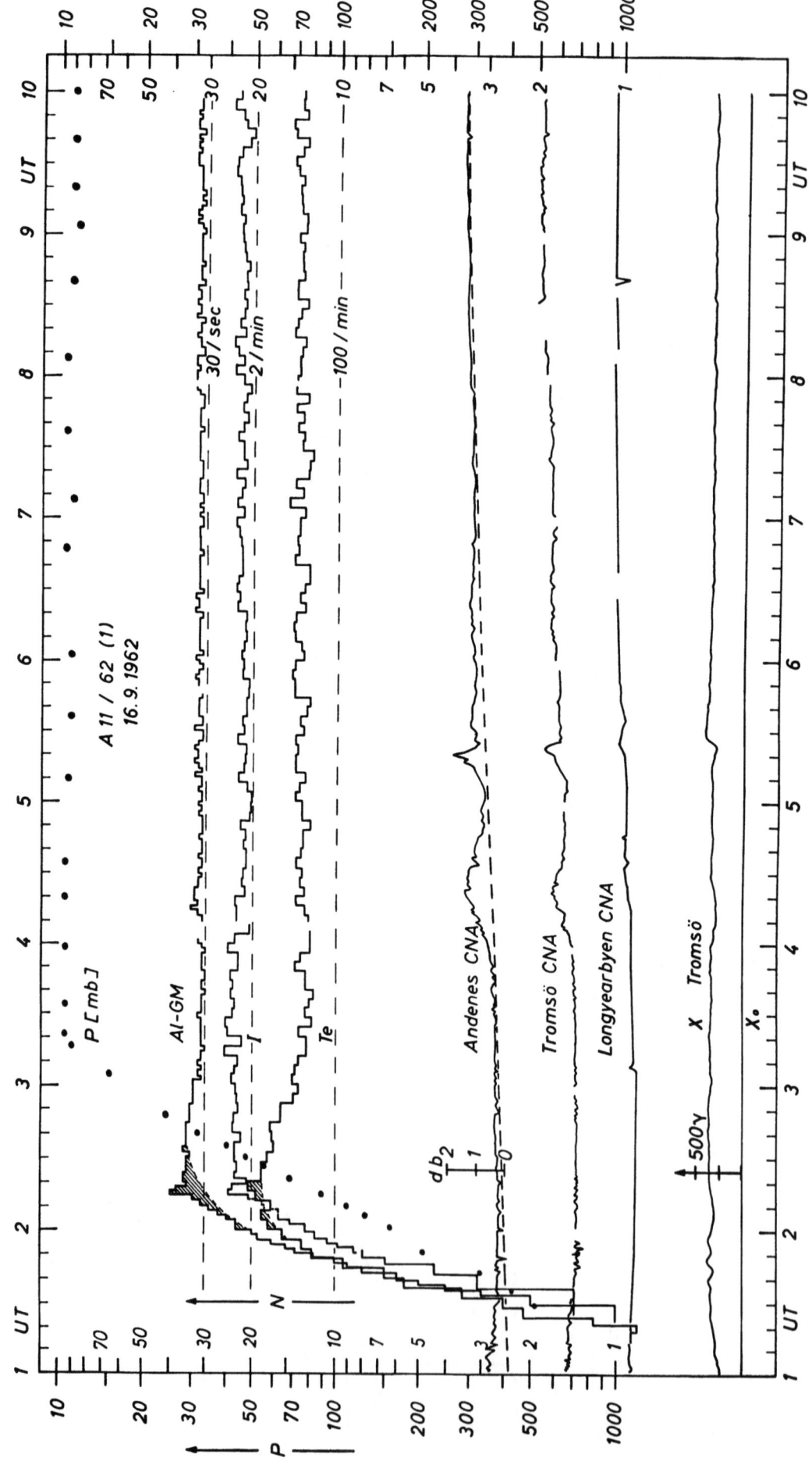

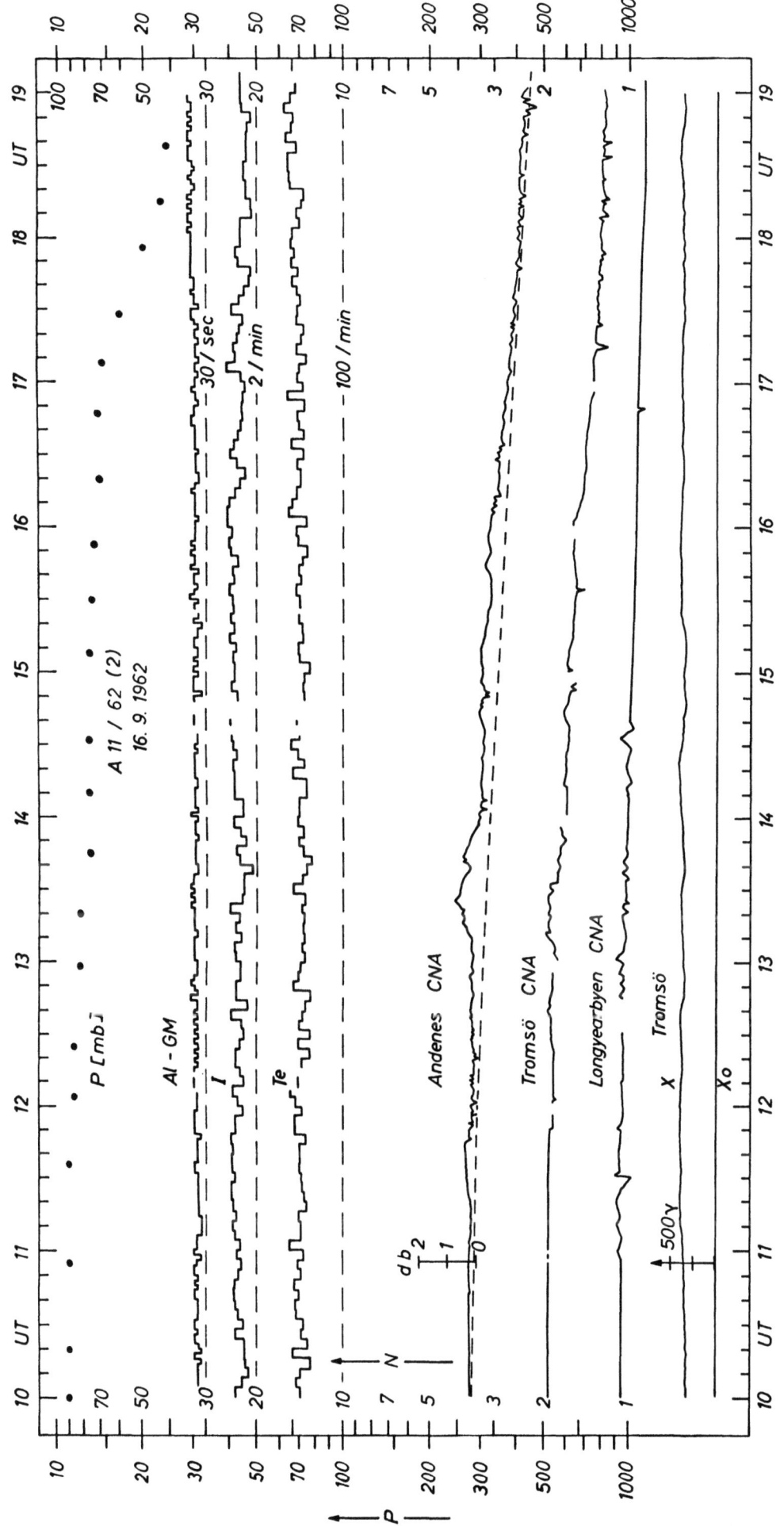

Balloon Flights 1963

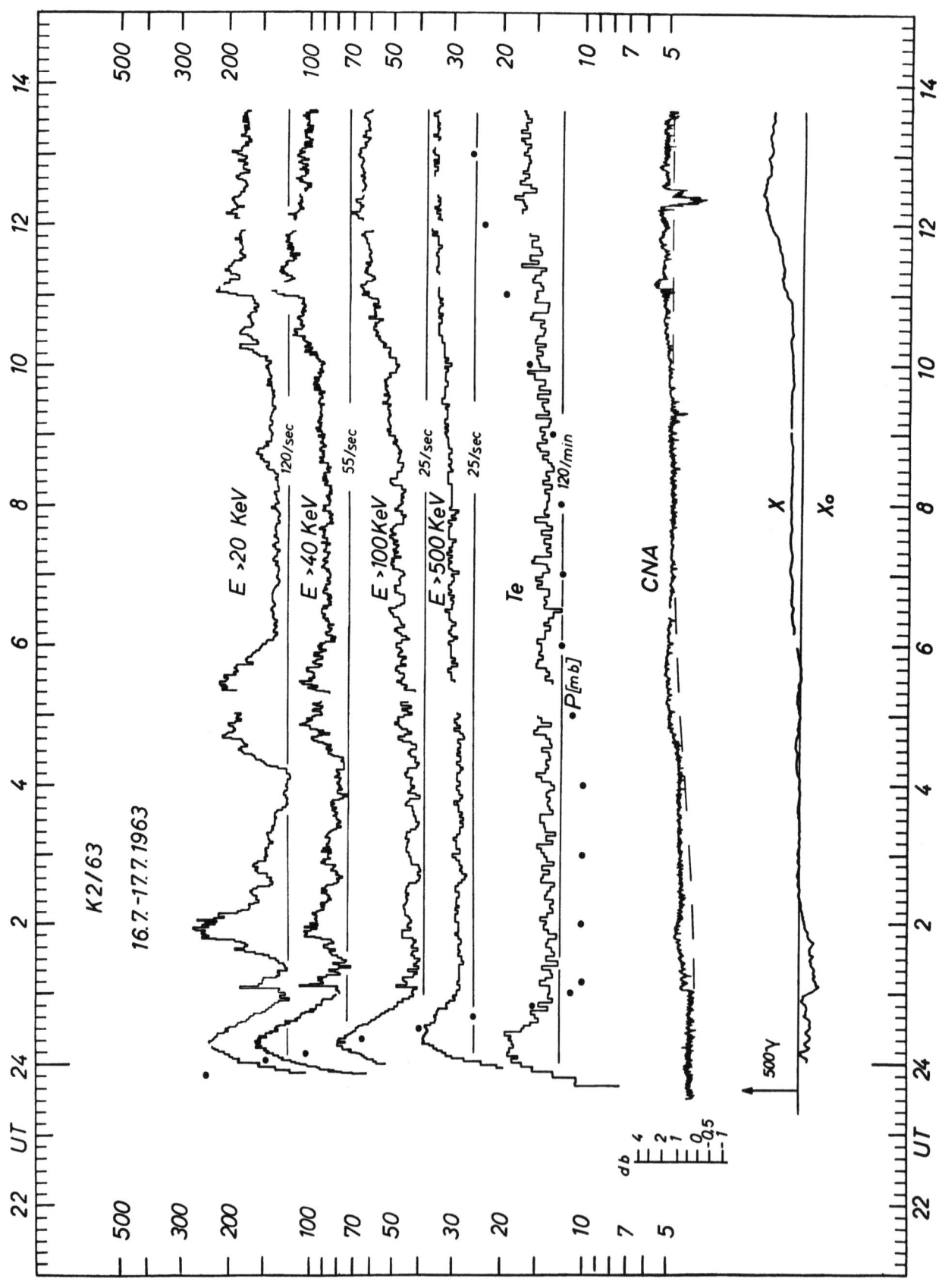

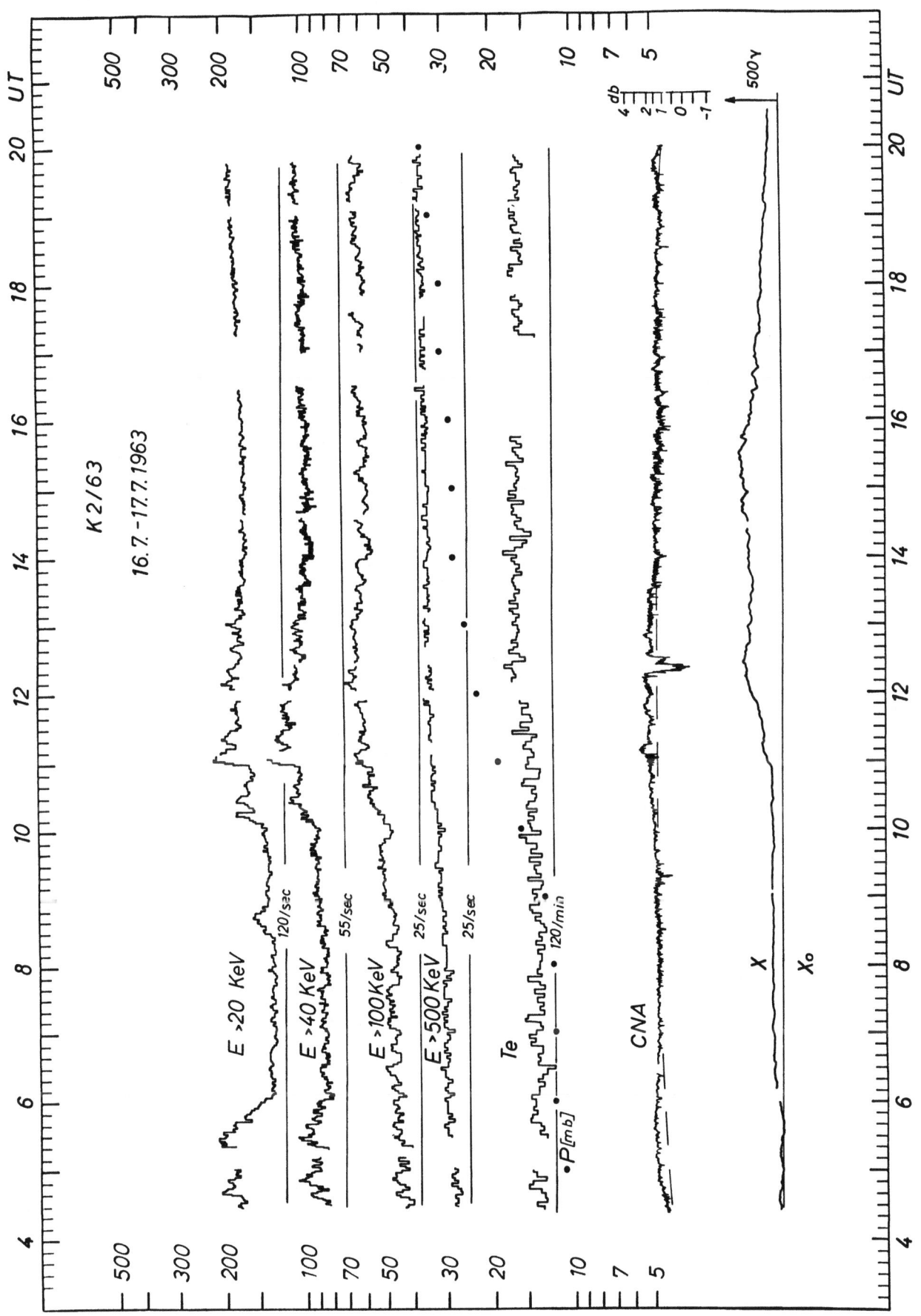

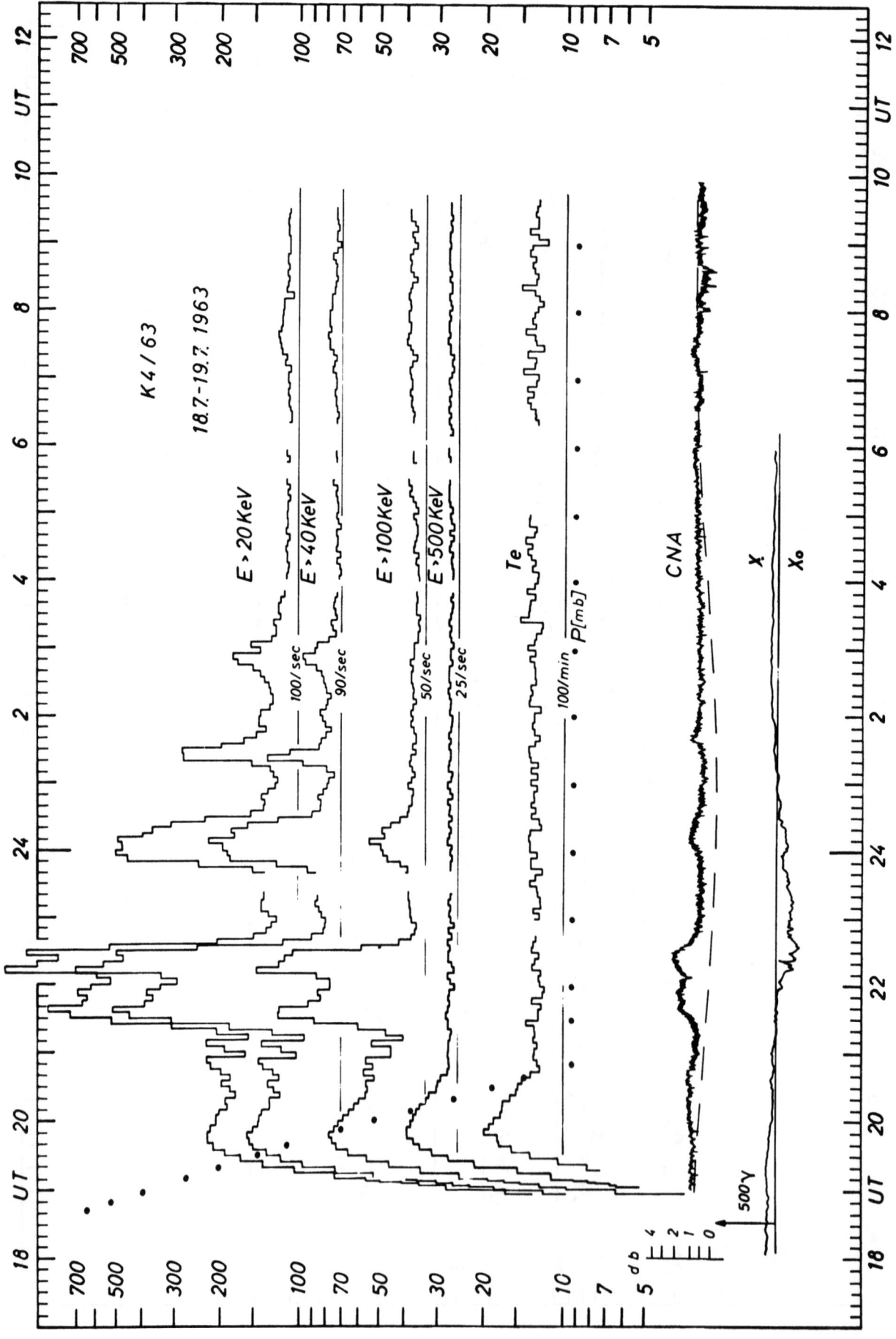

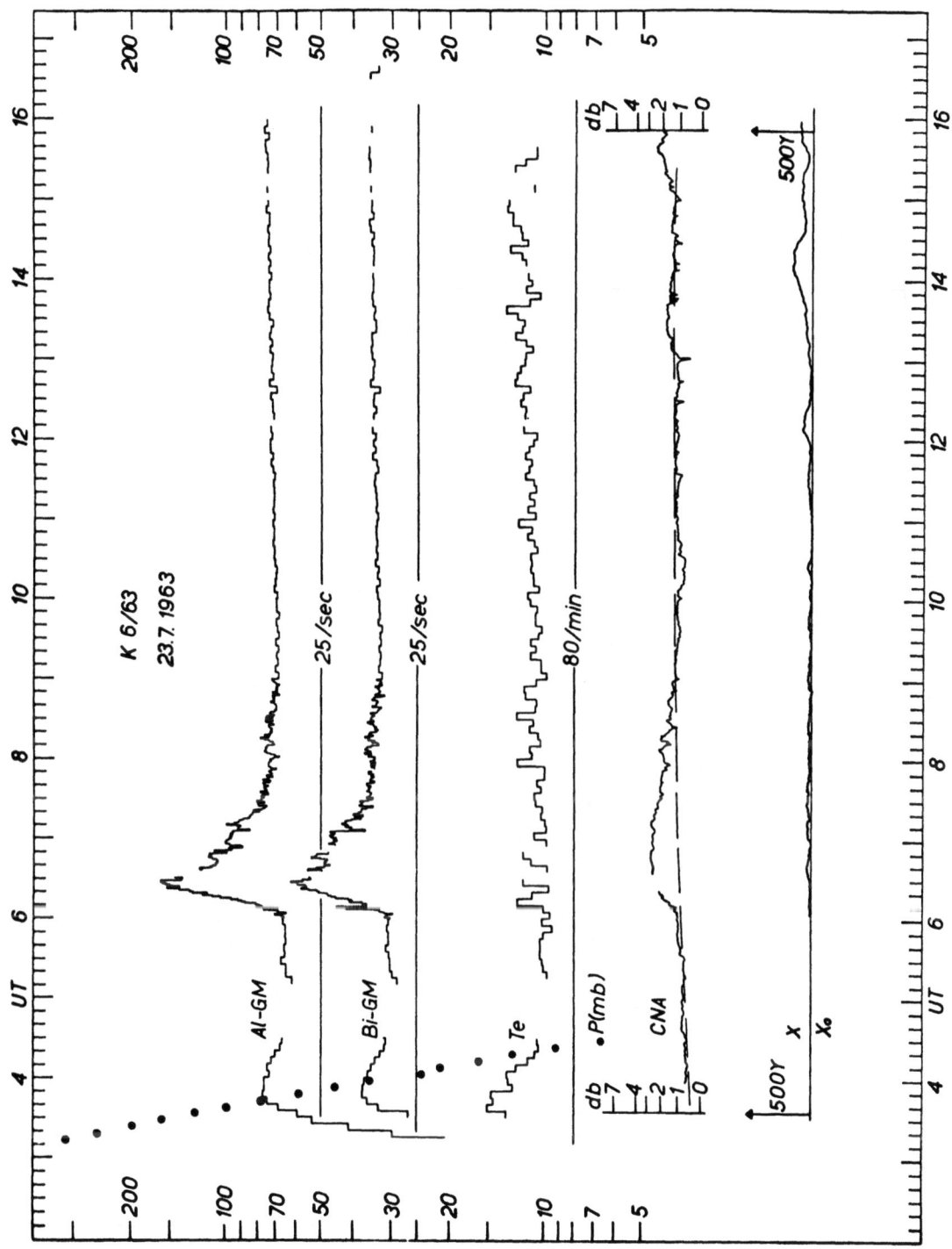

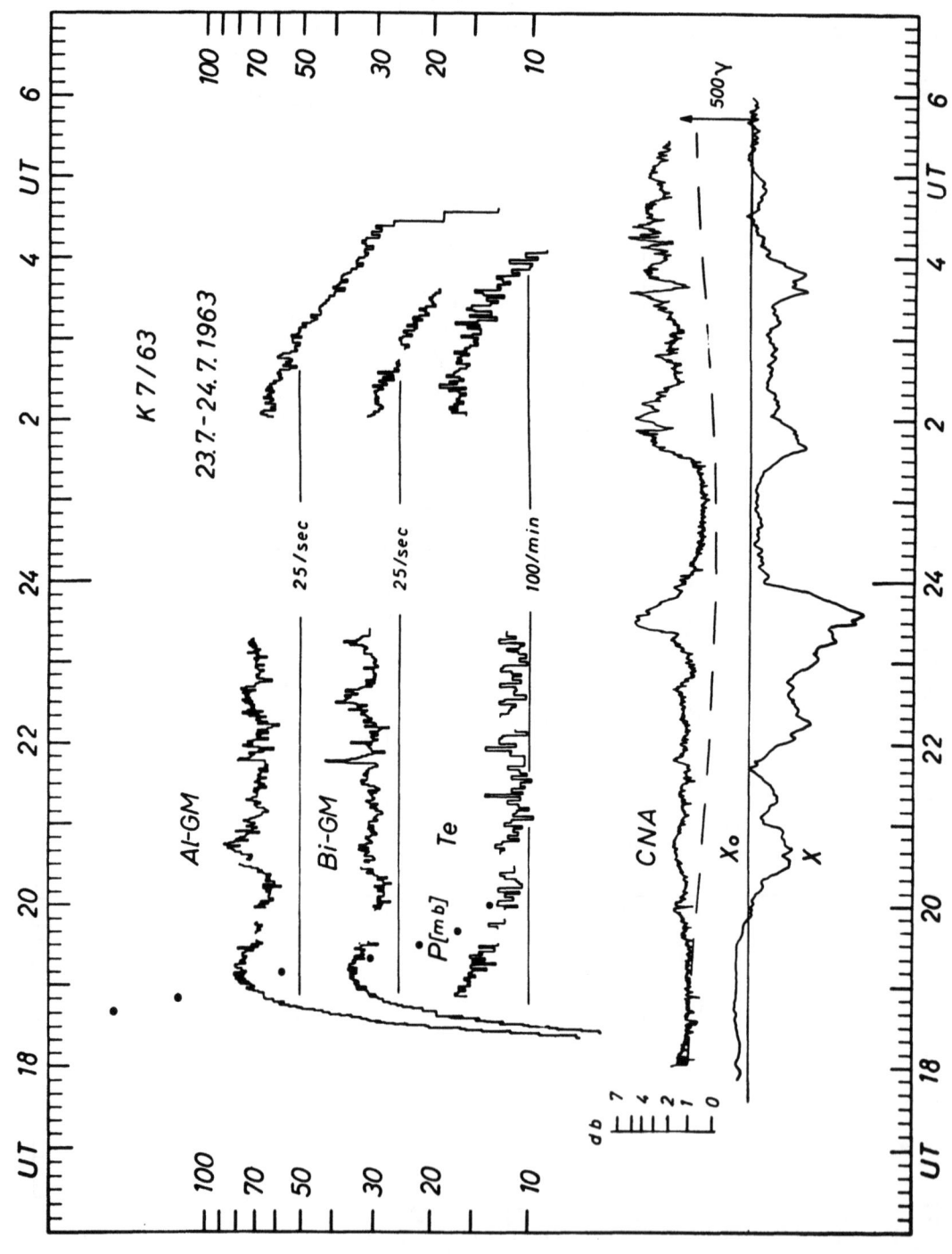

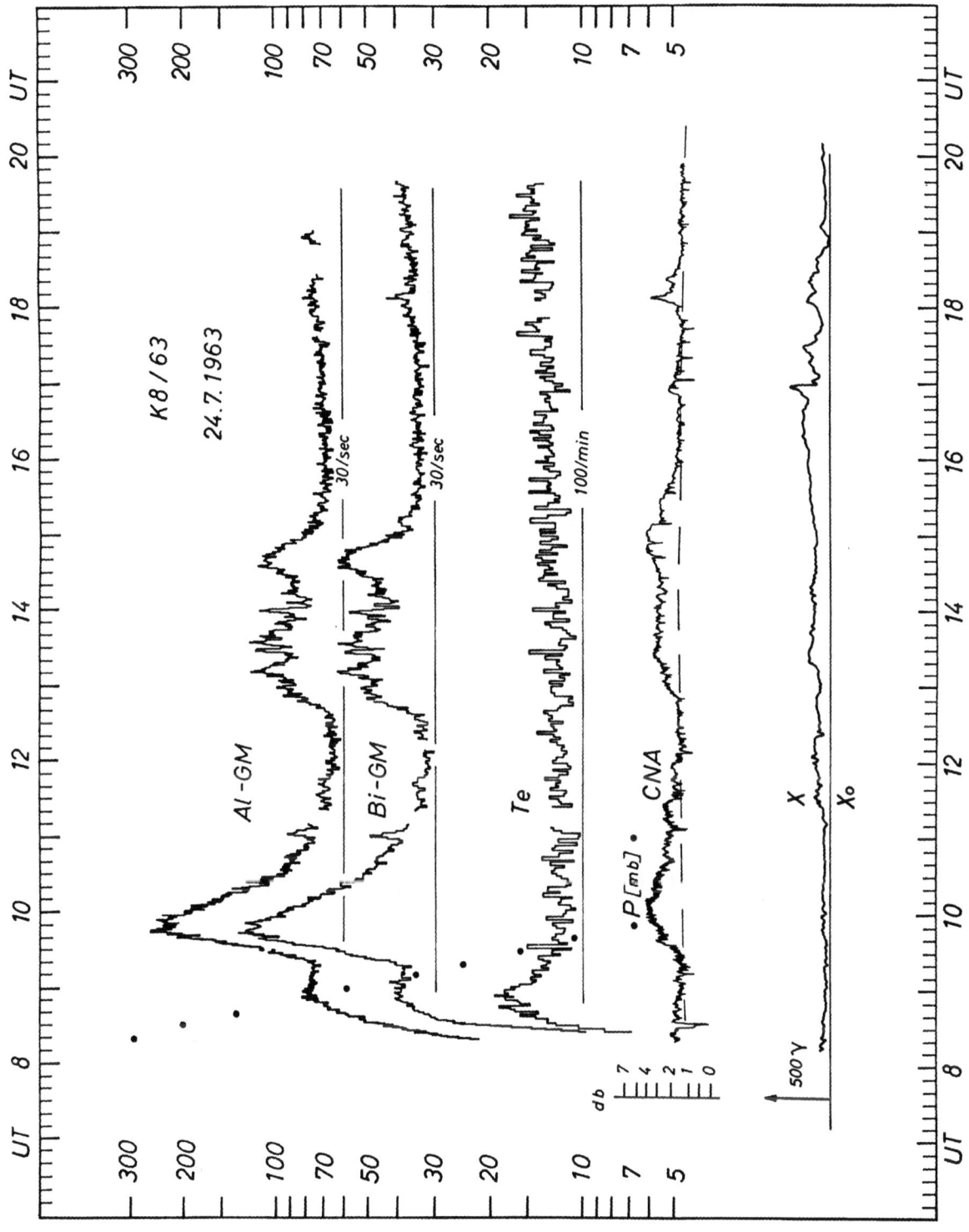

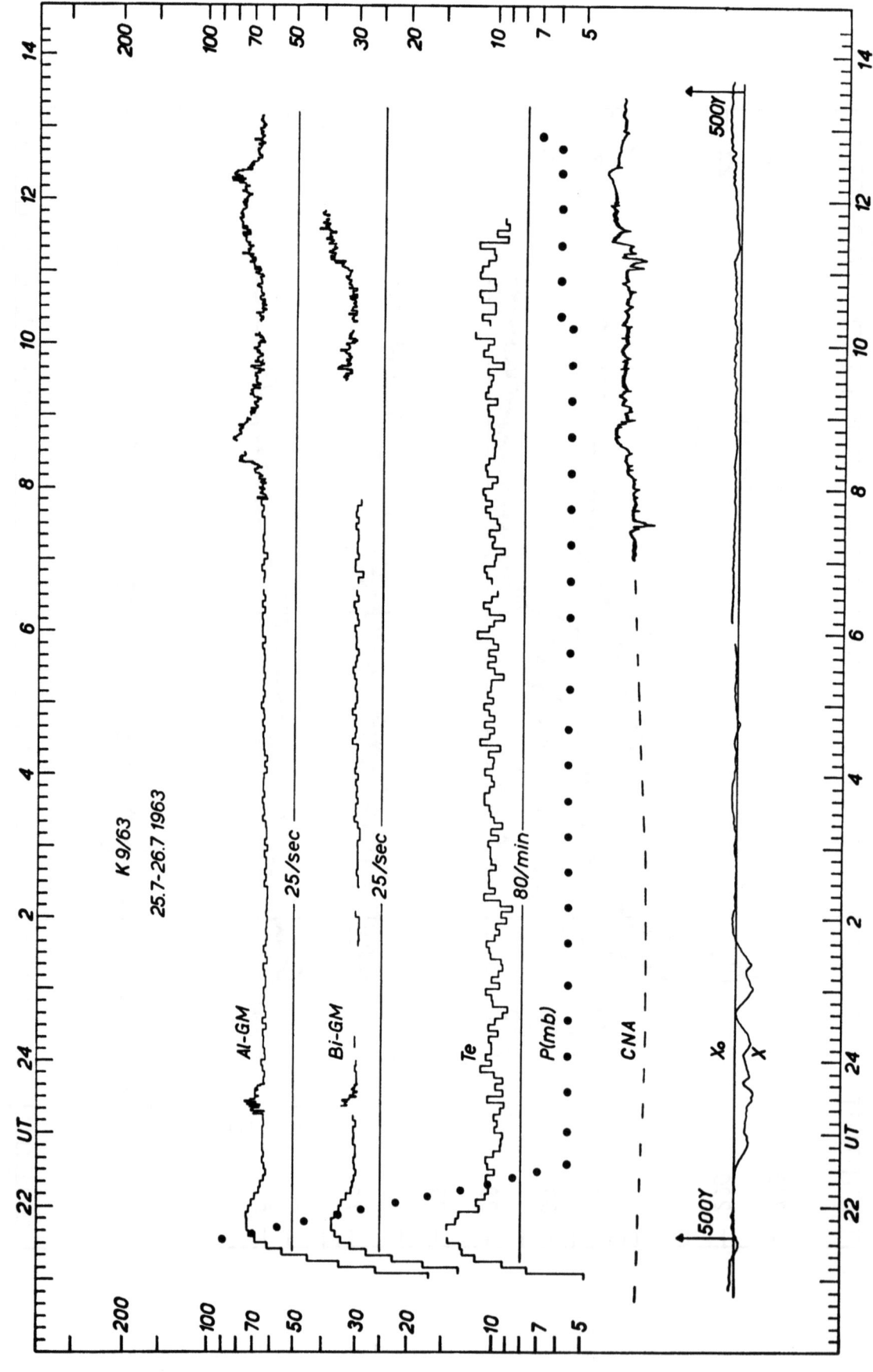

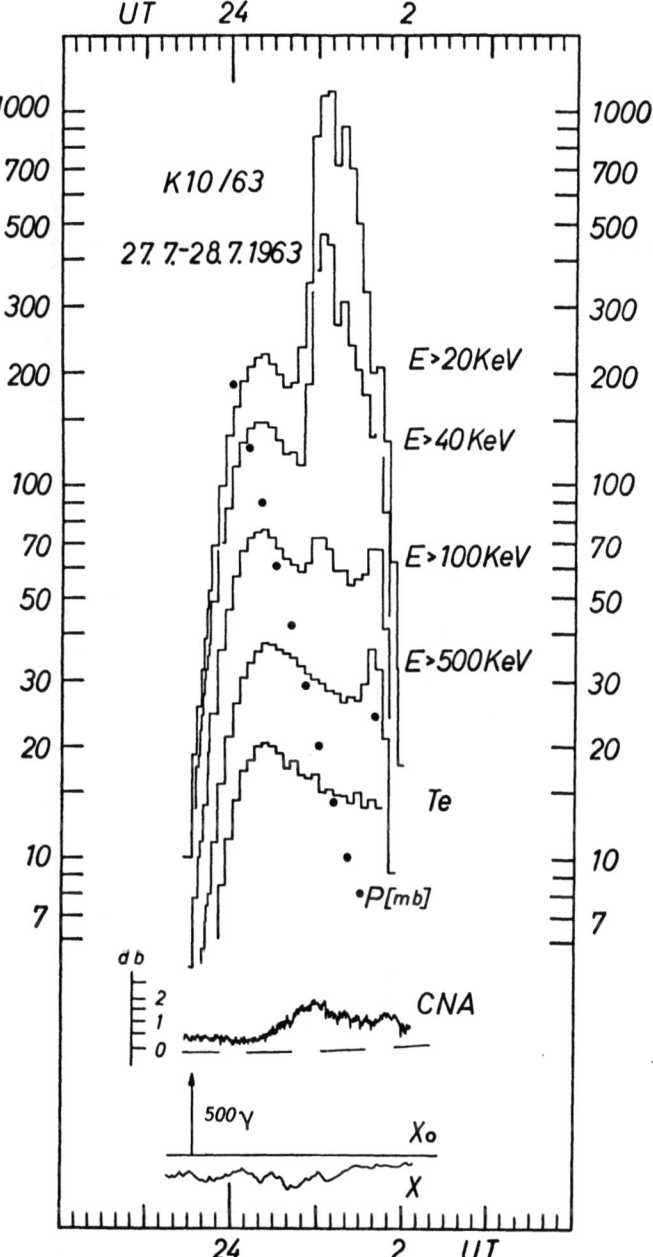

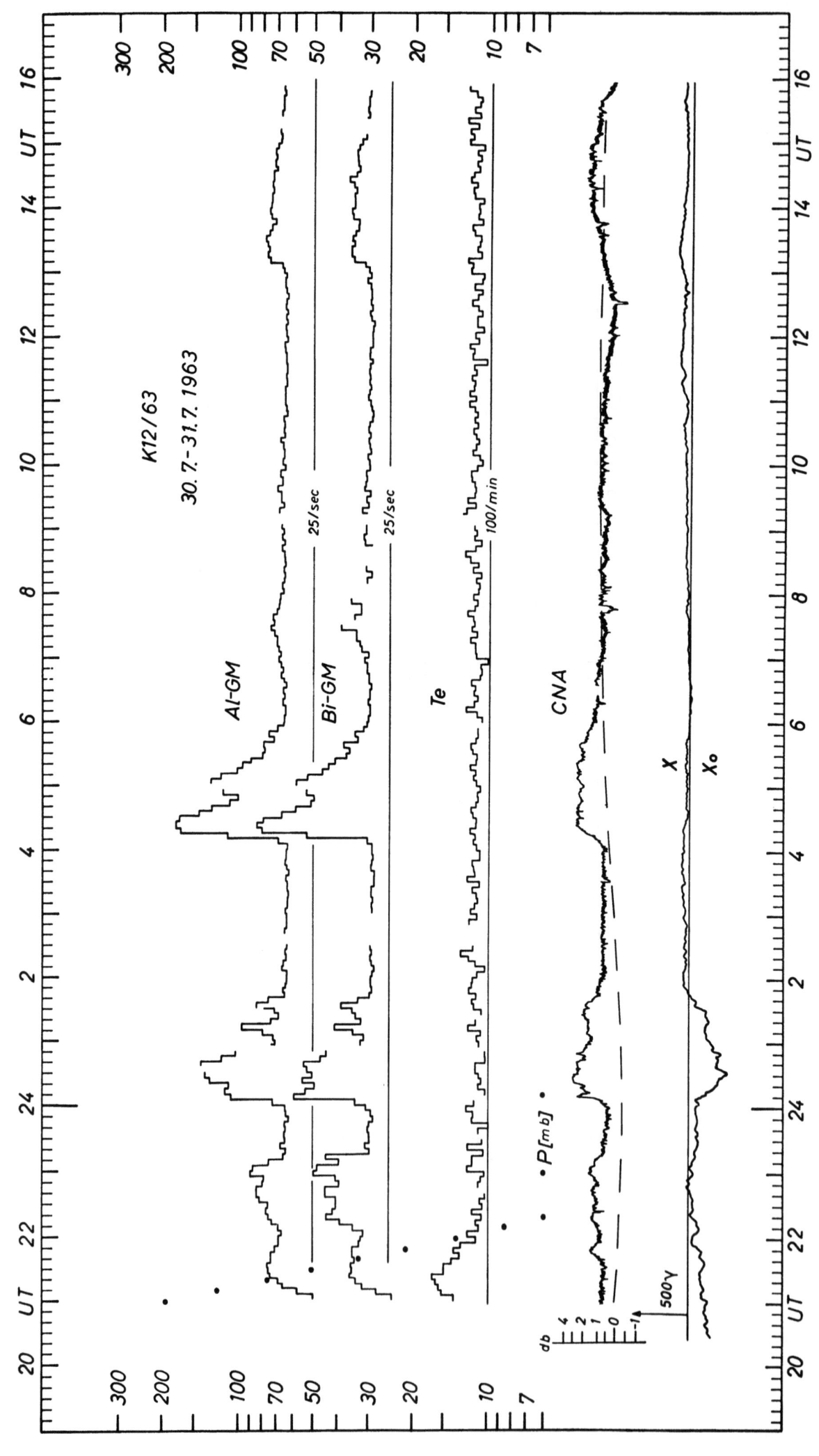

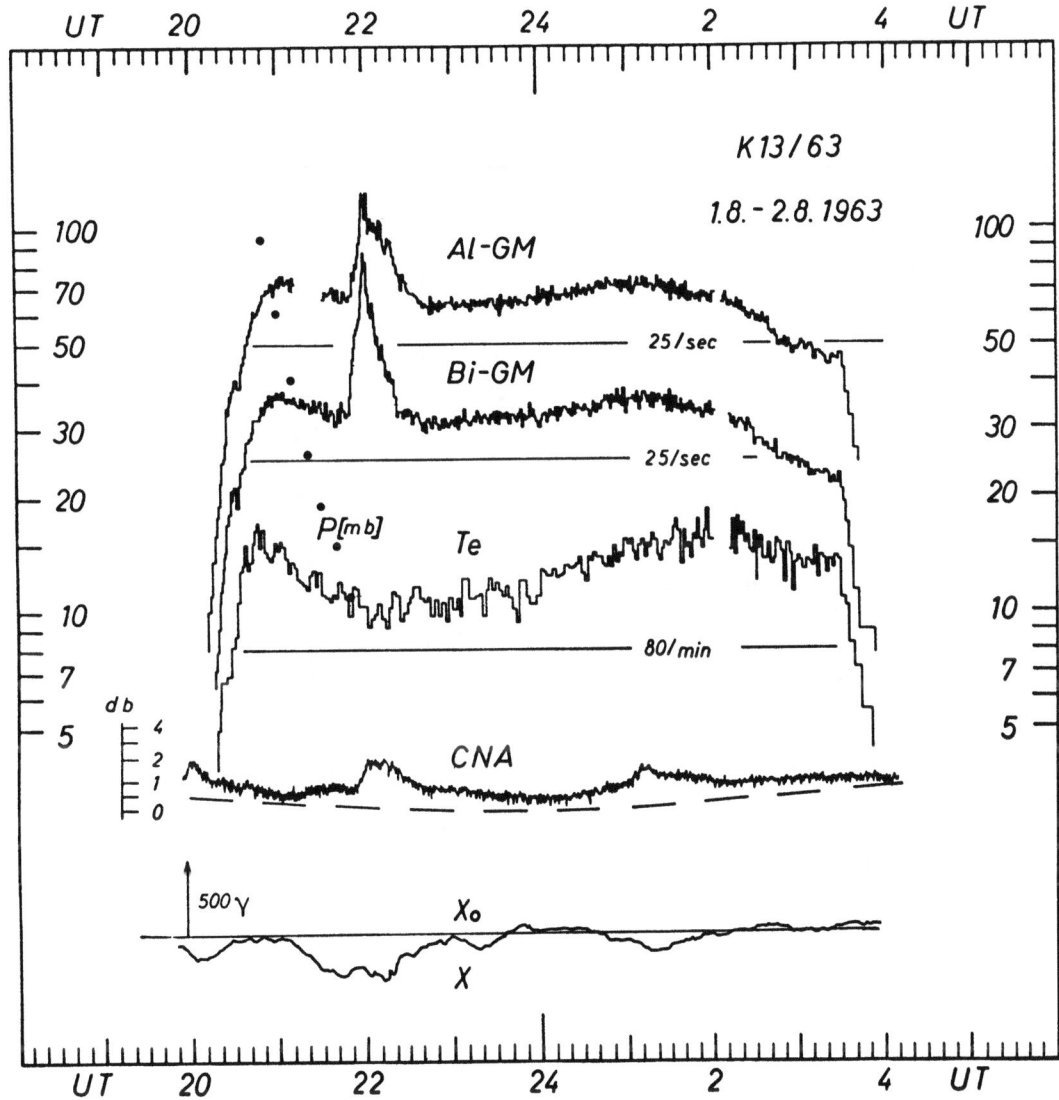

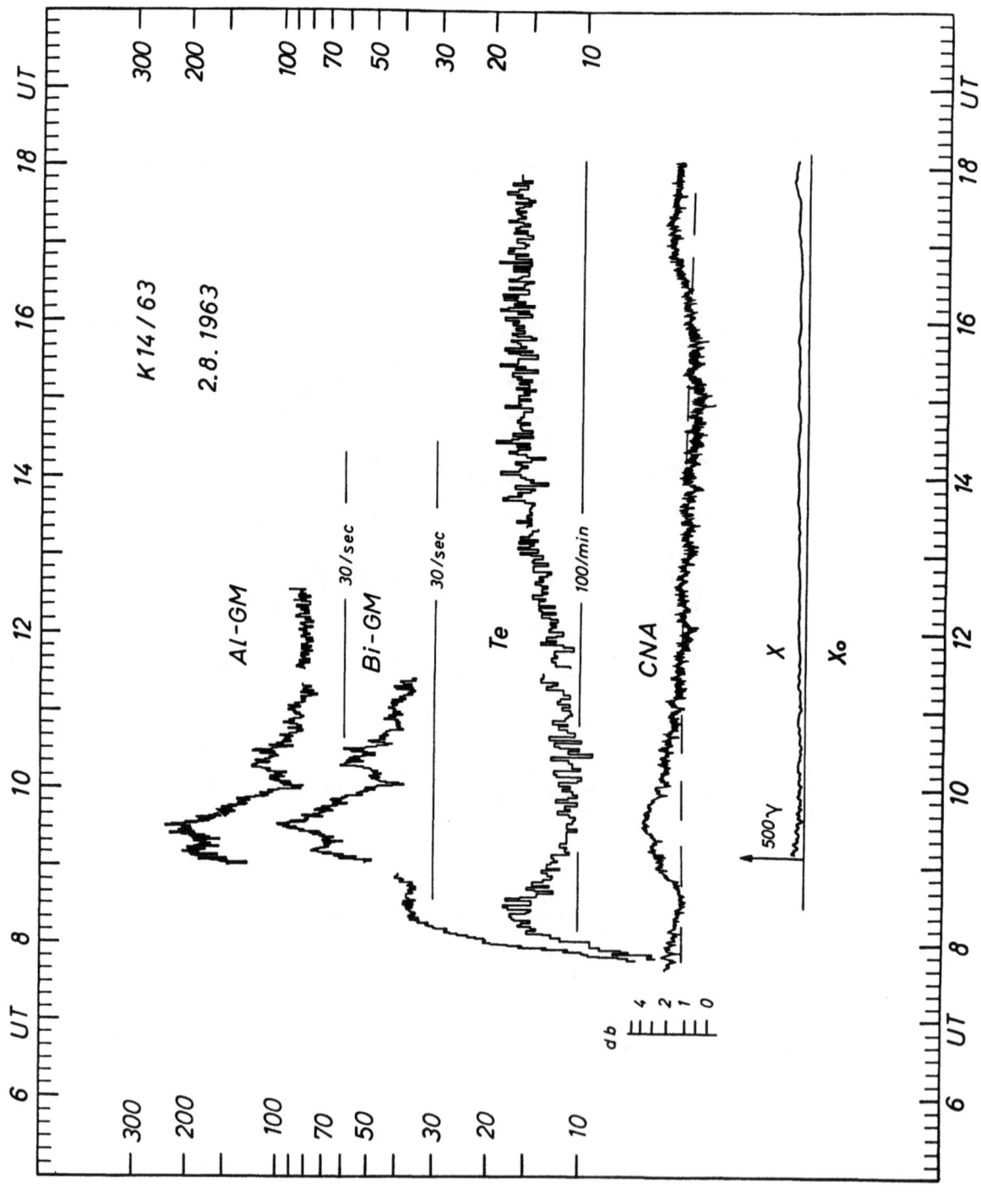

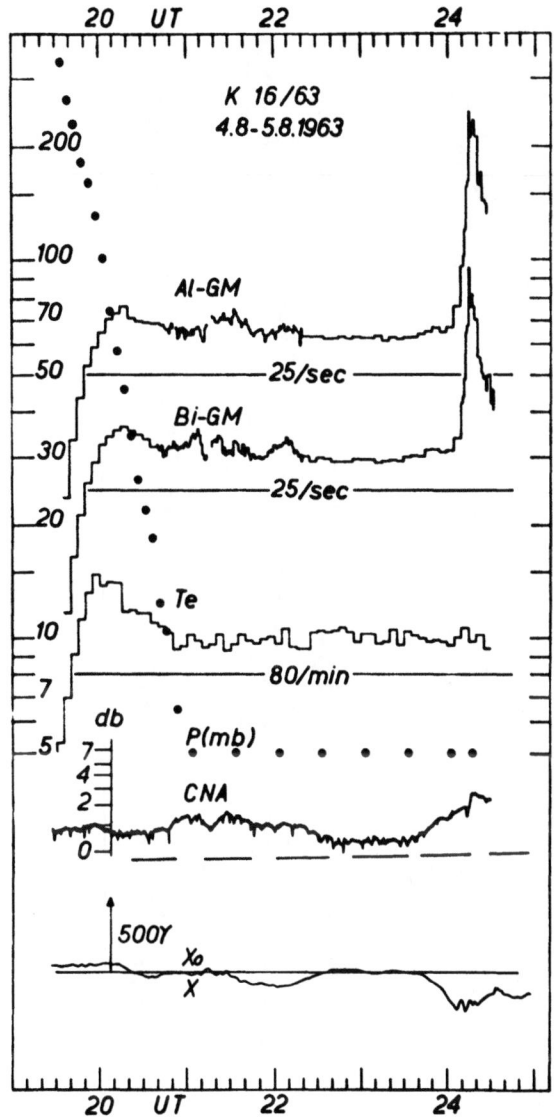

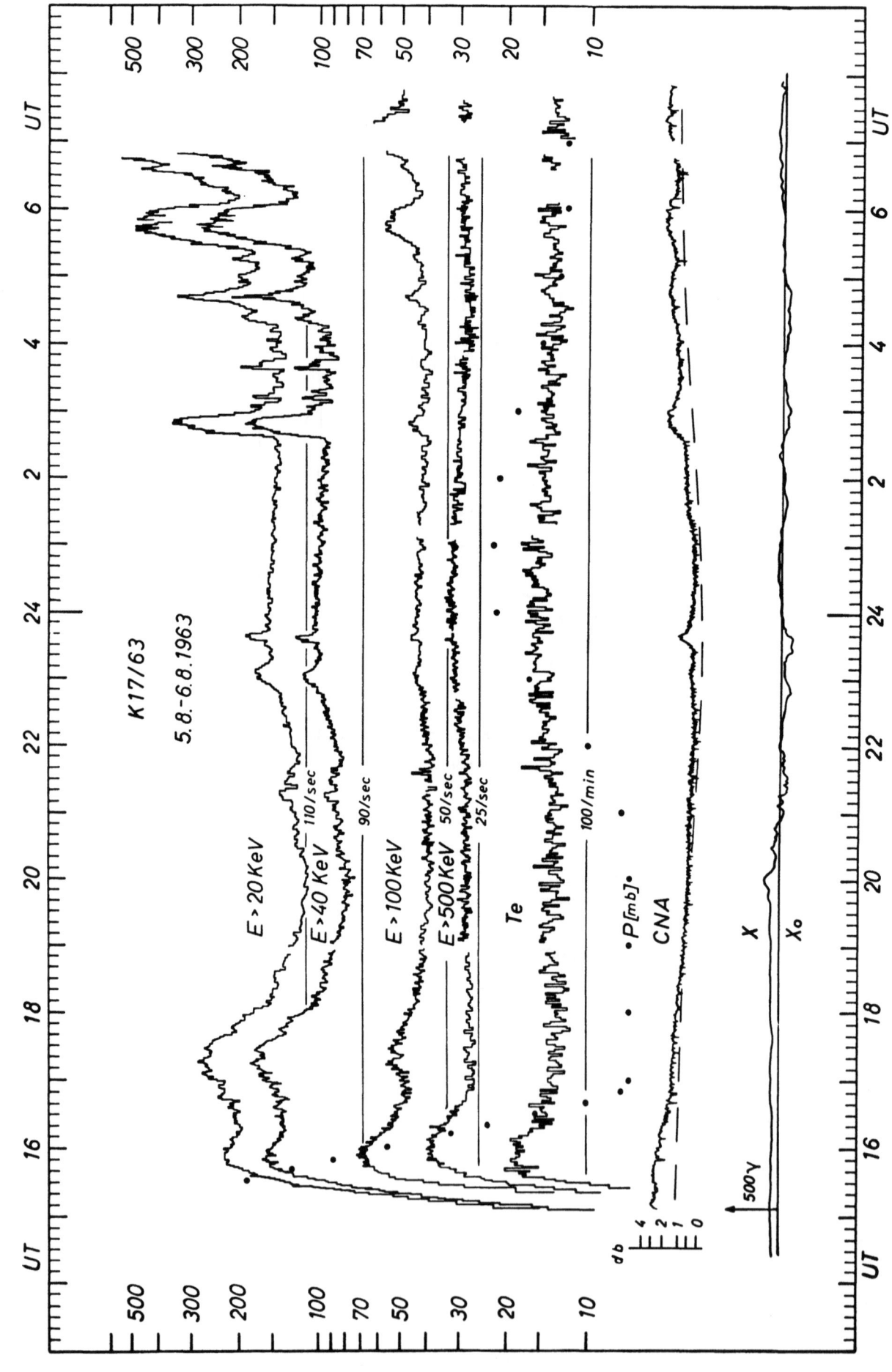

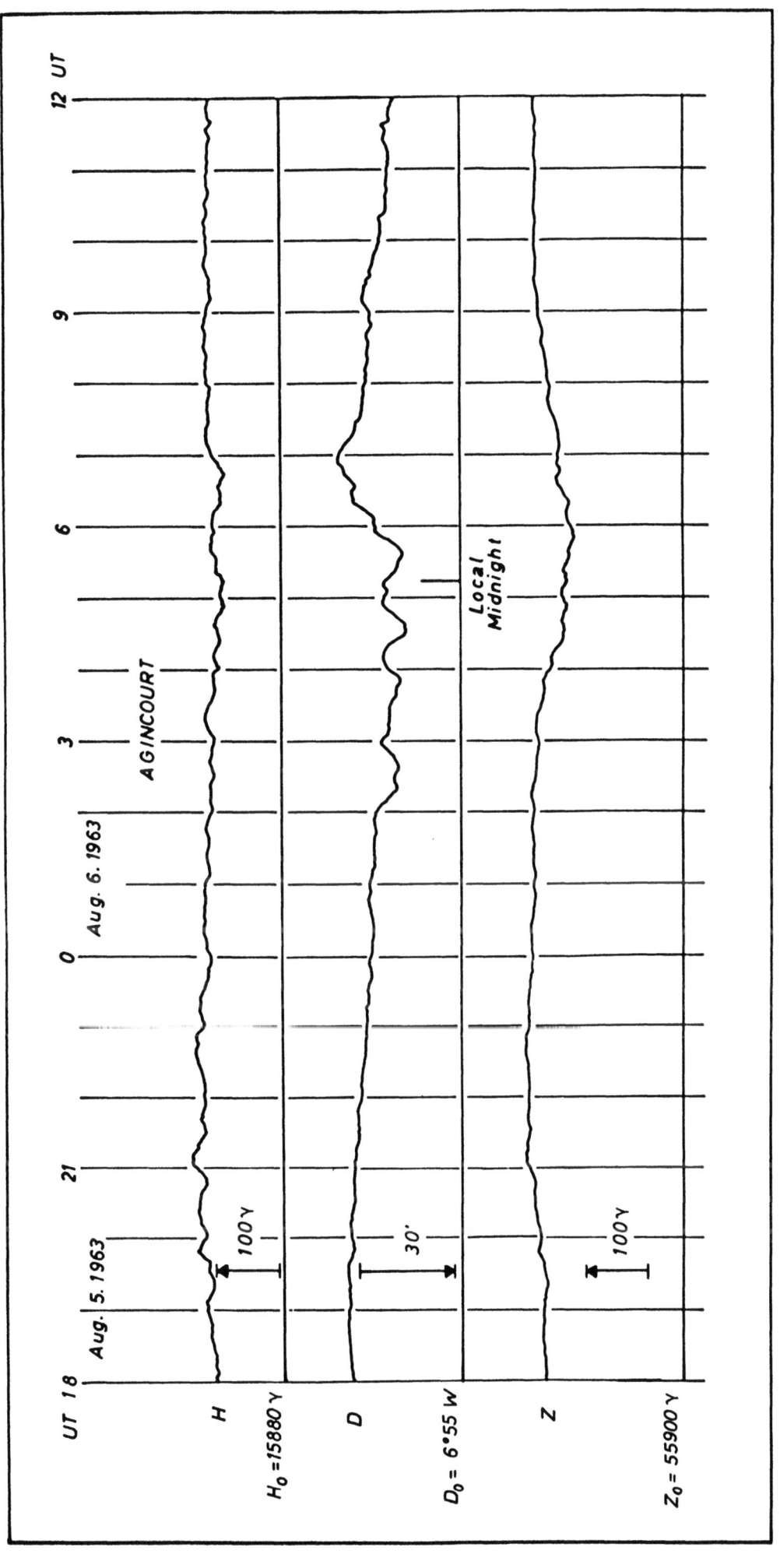

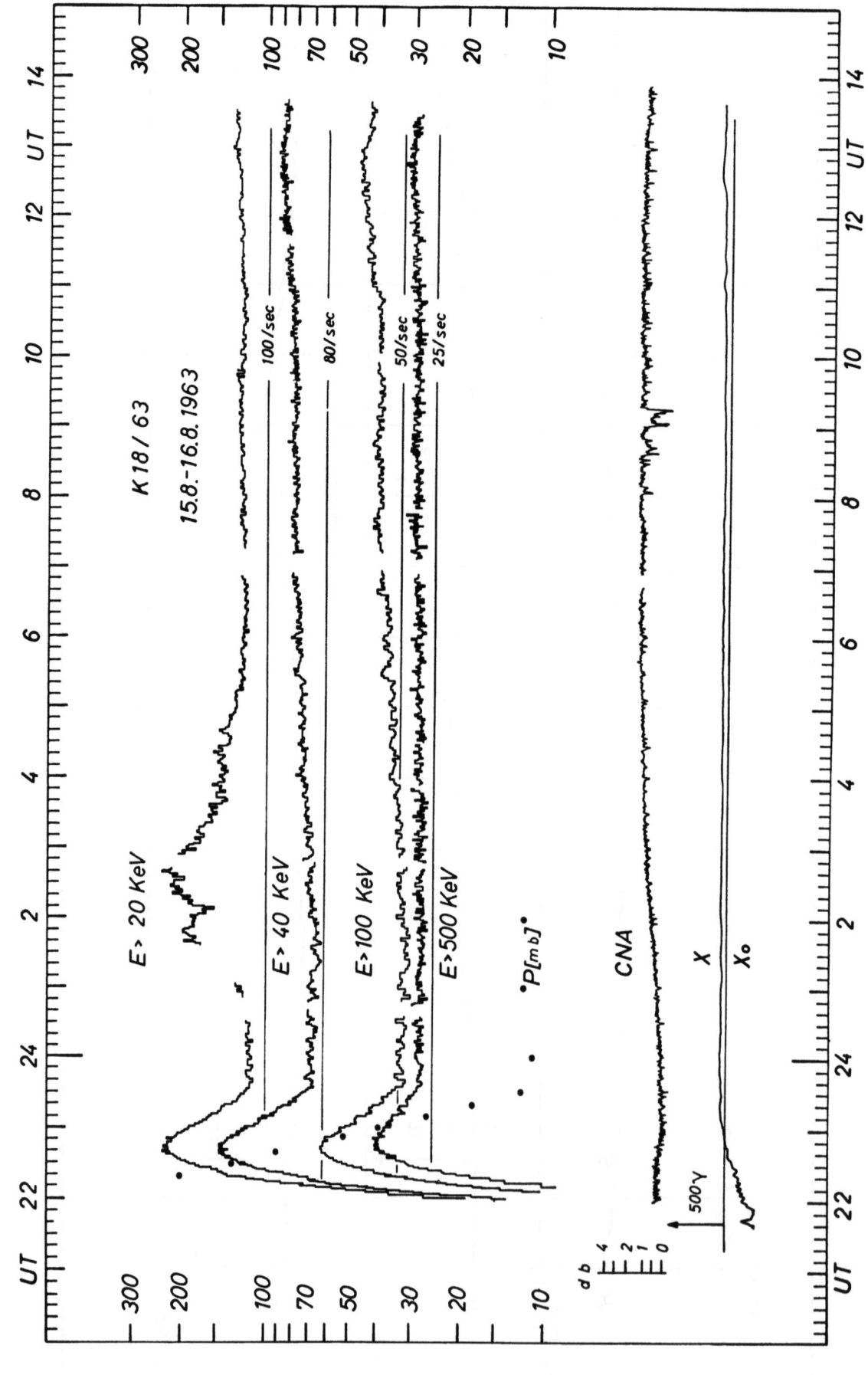

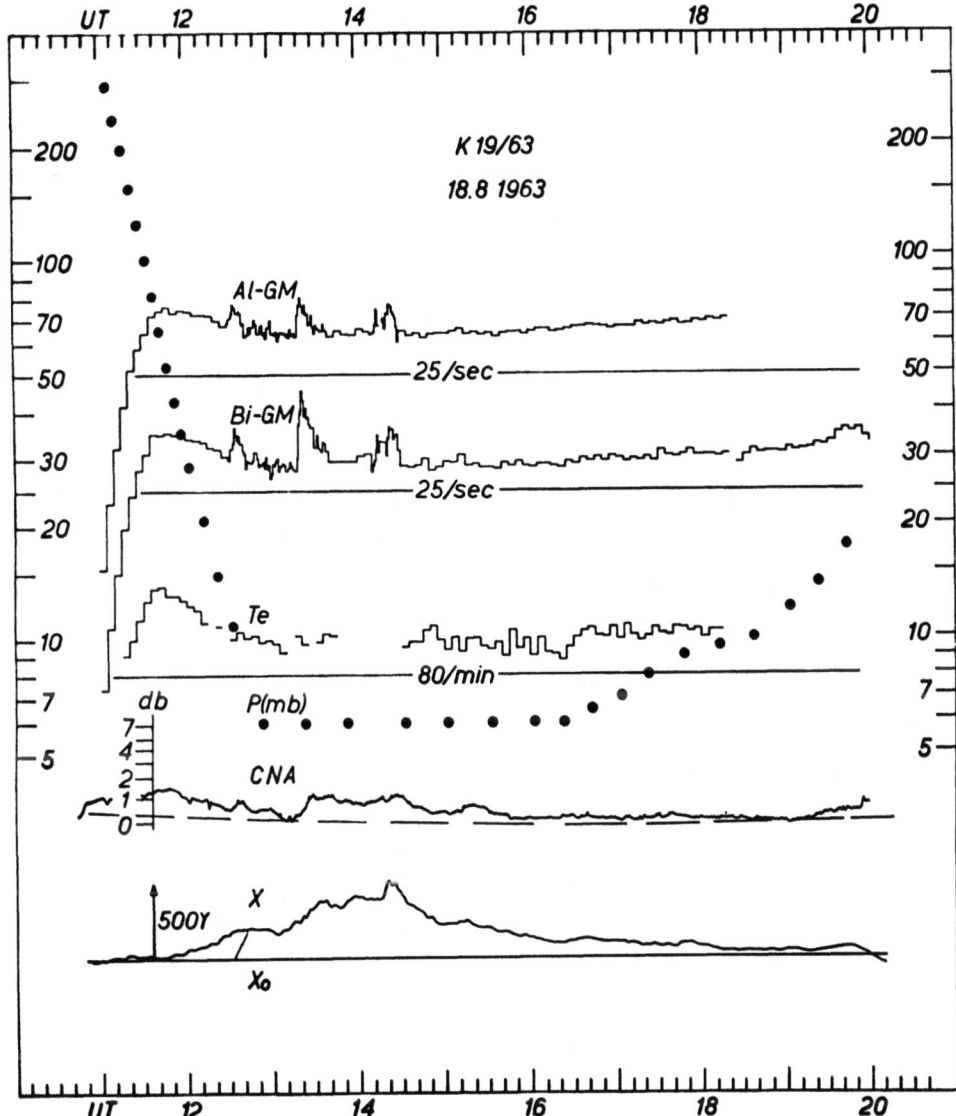

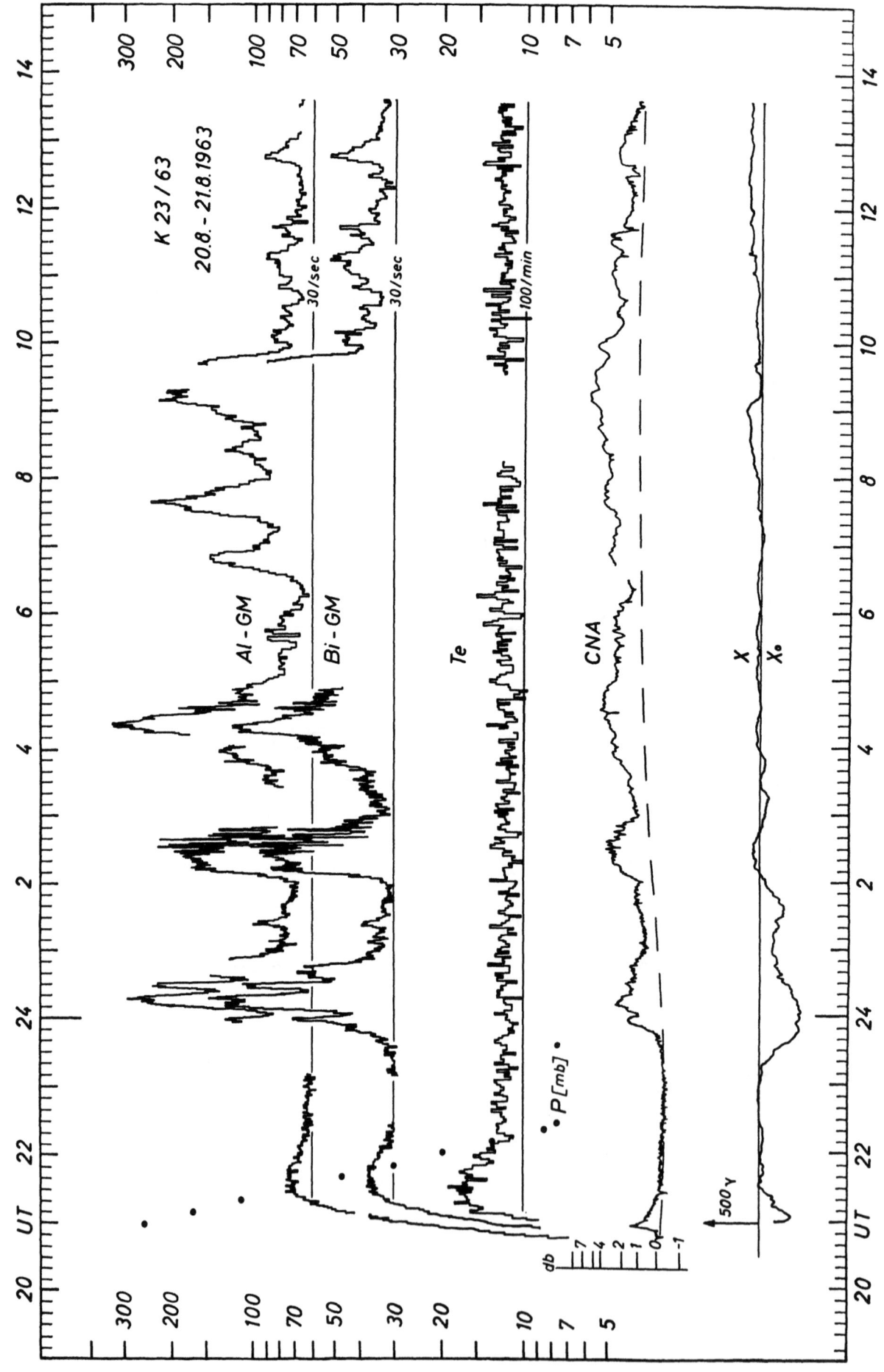

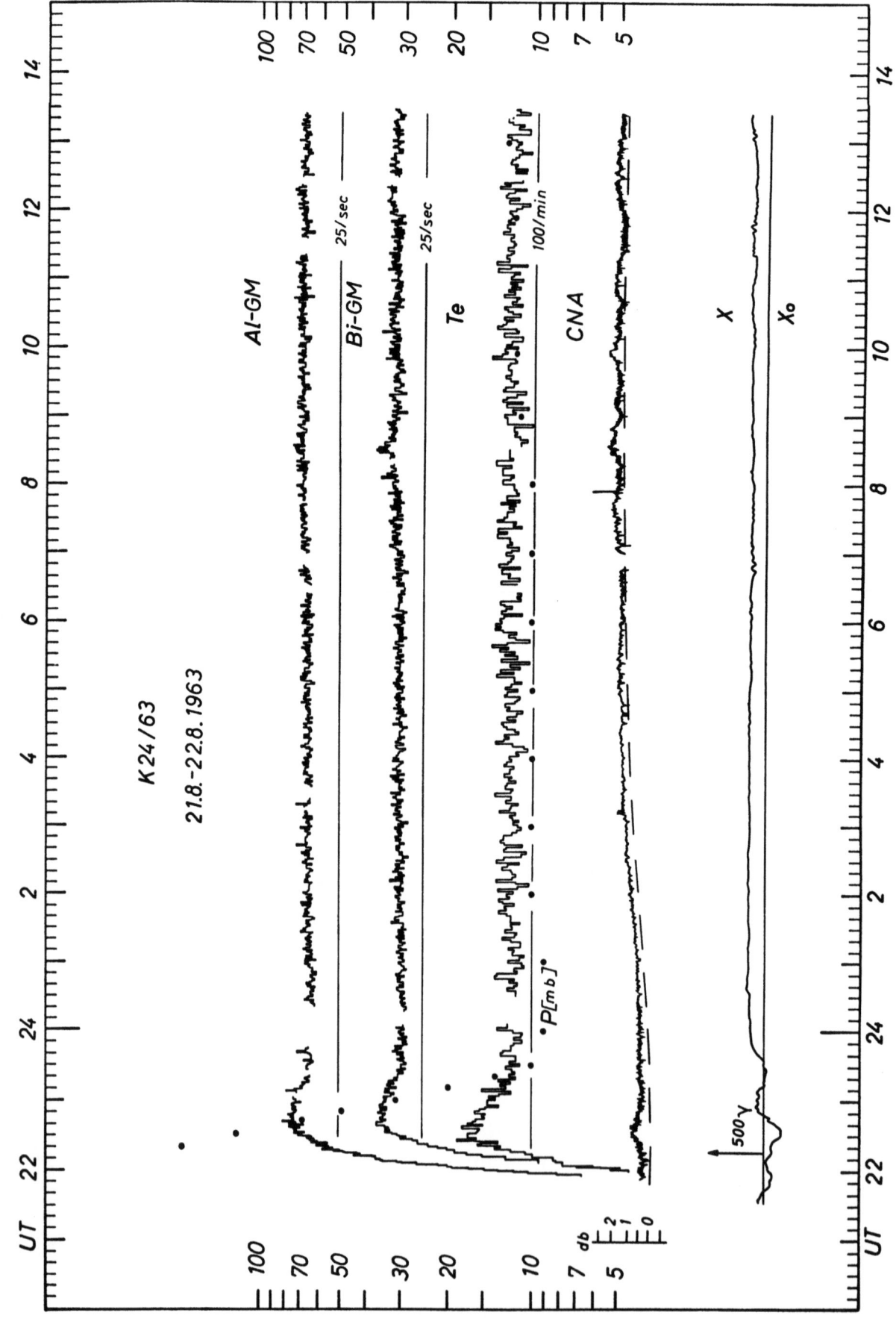

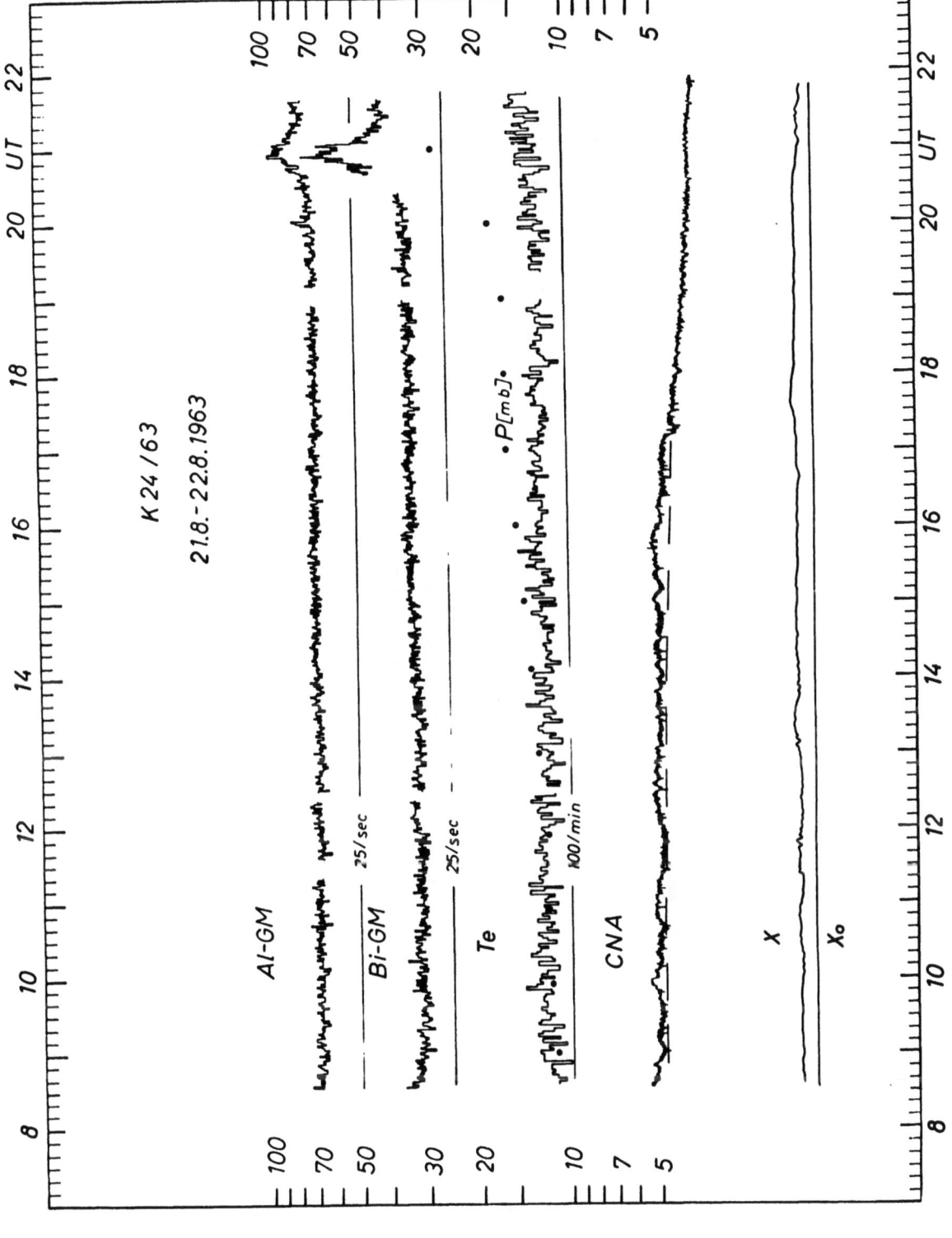

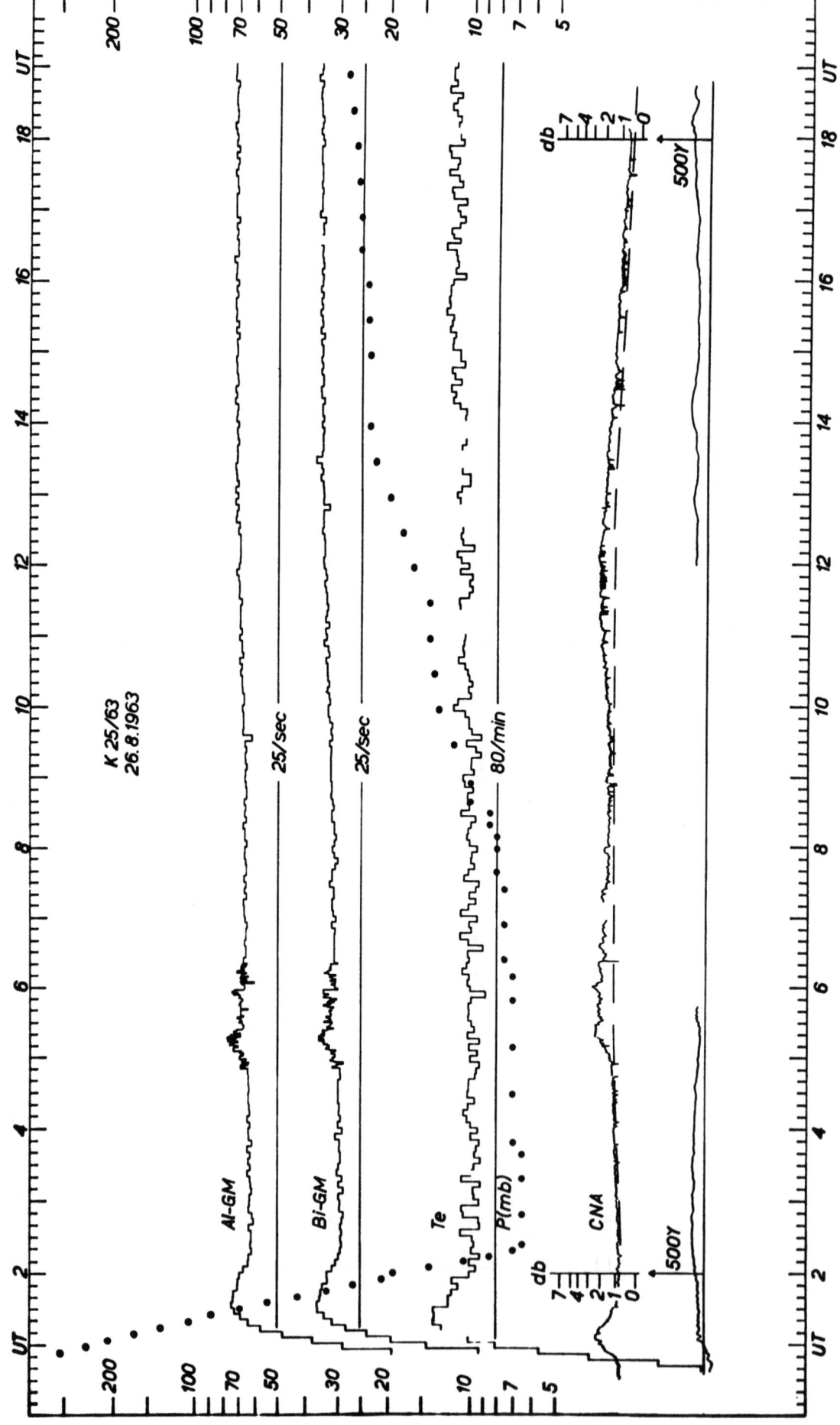

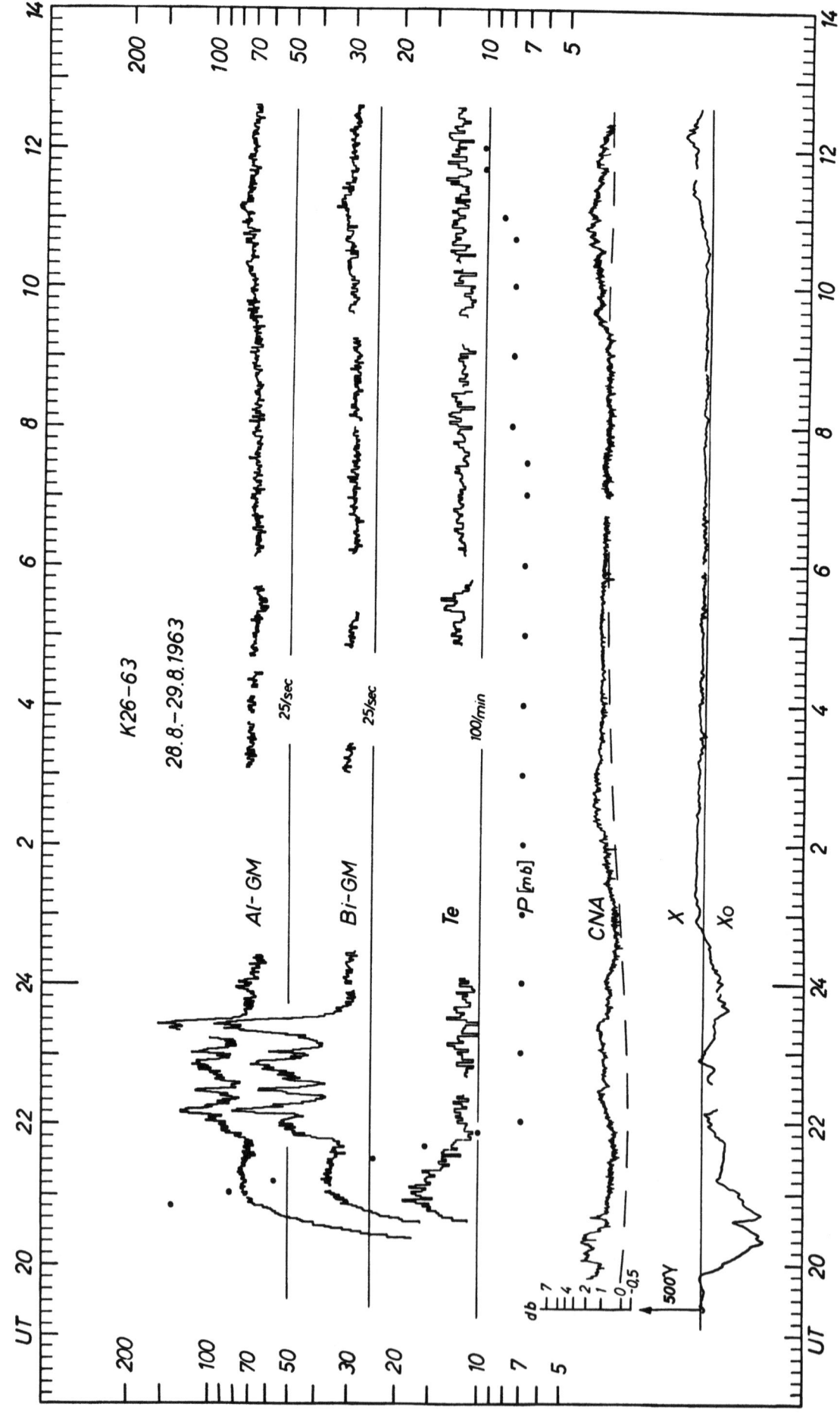

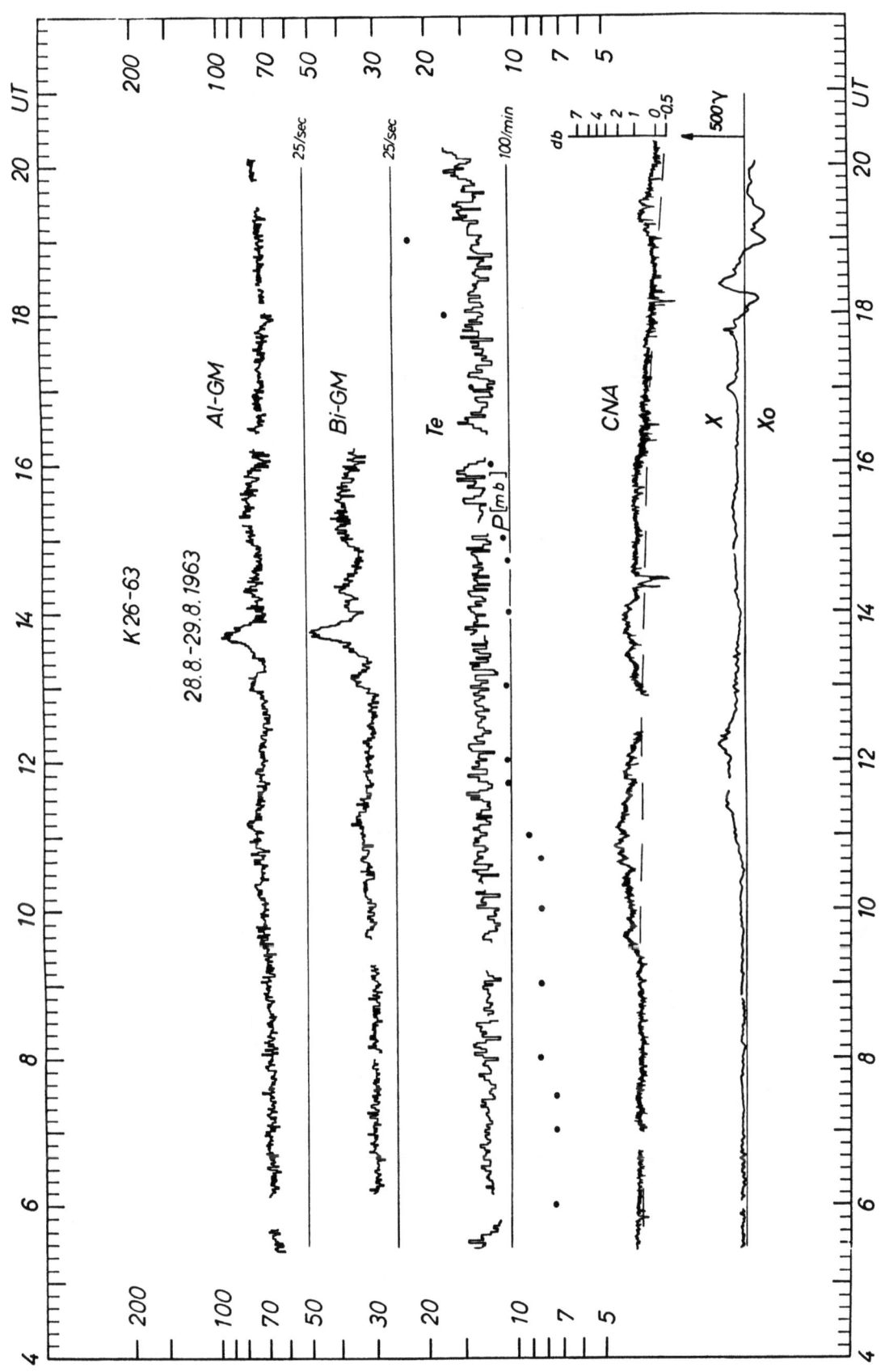

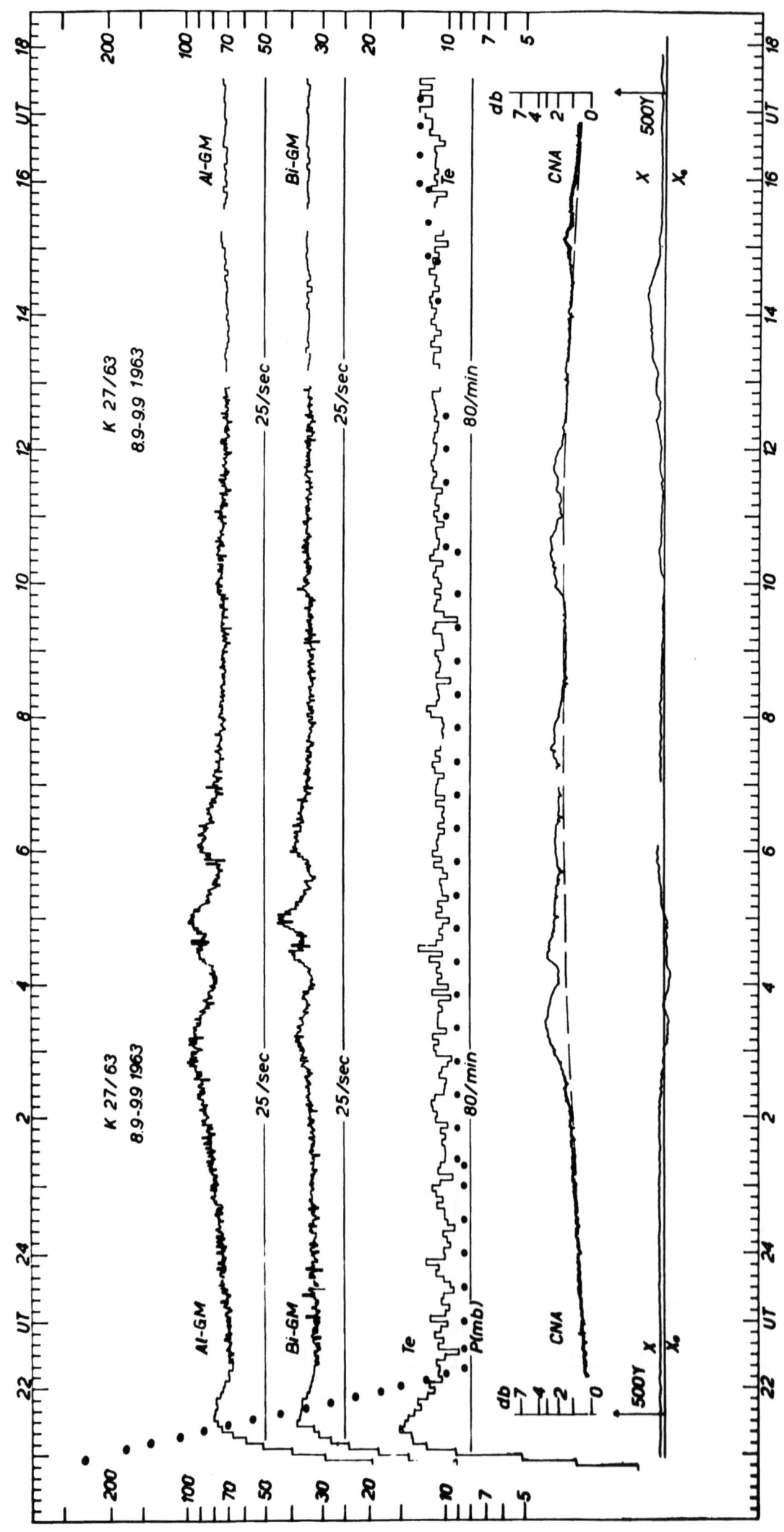

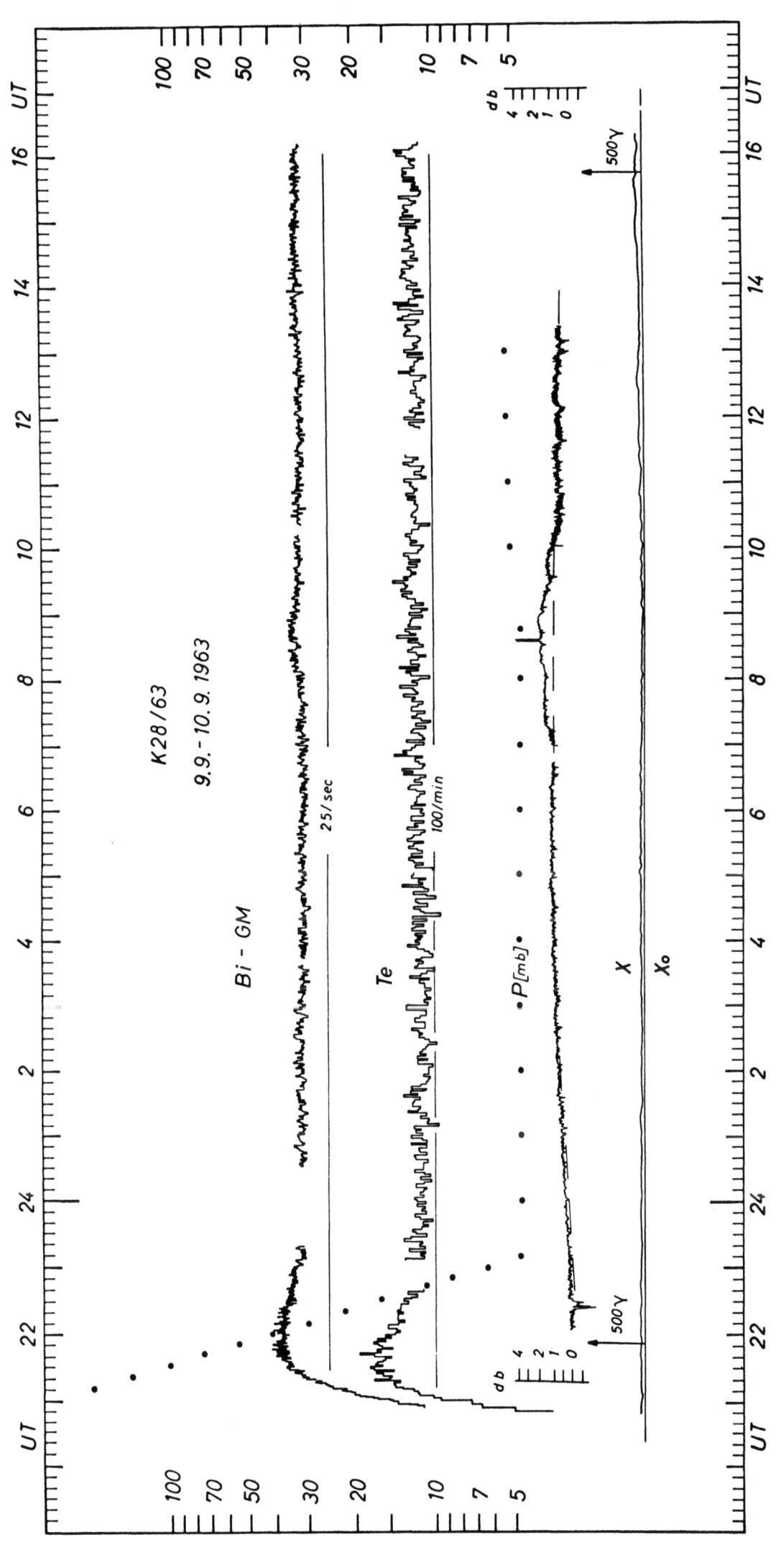

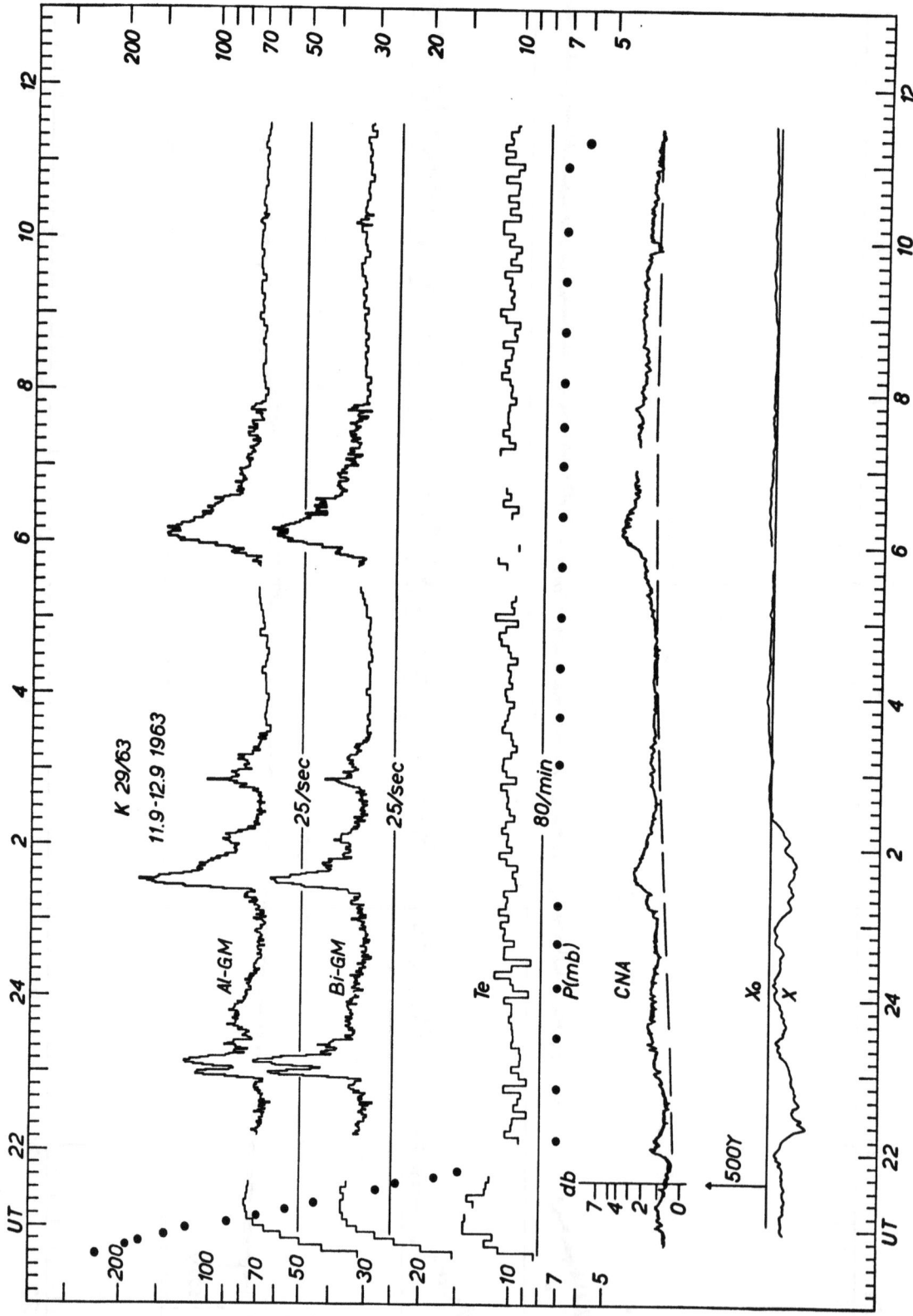

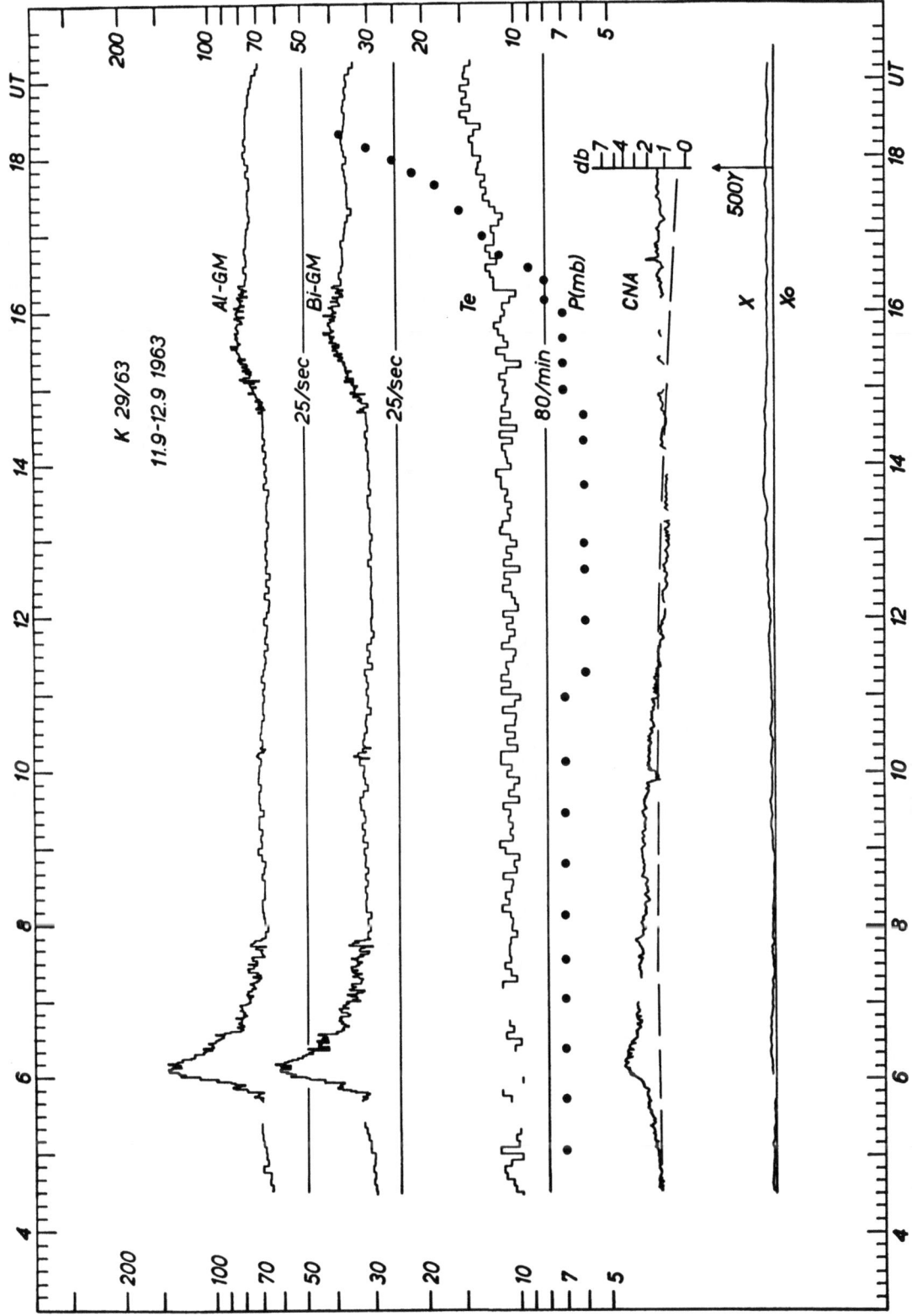

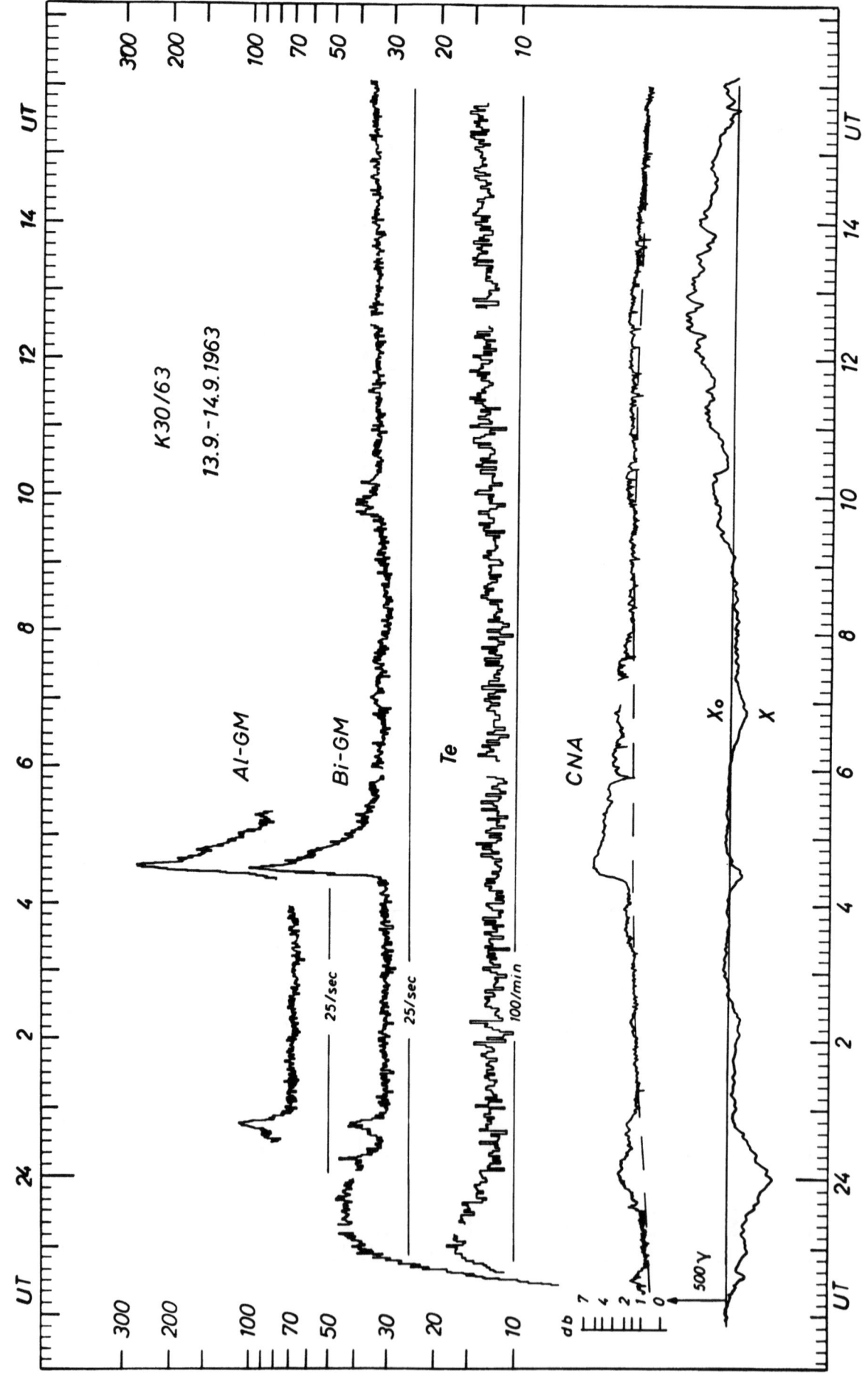

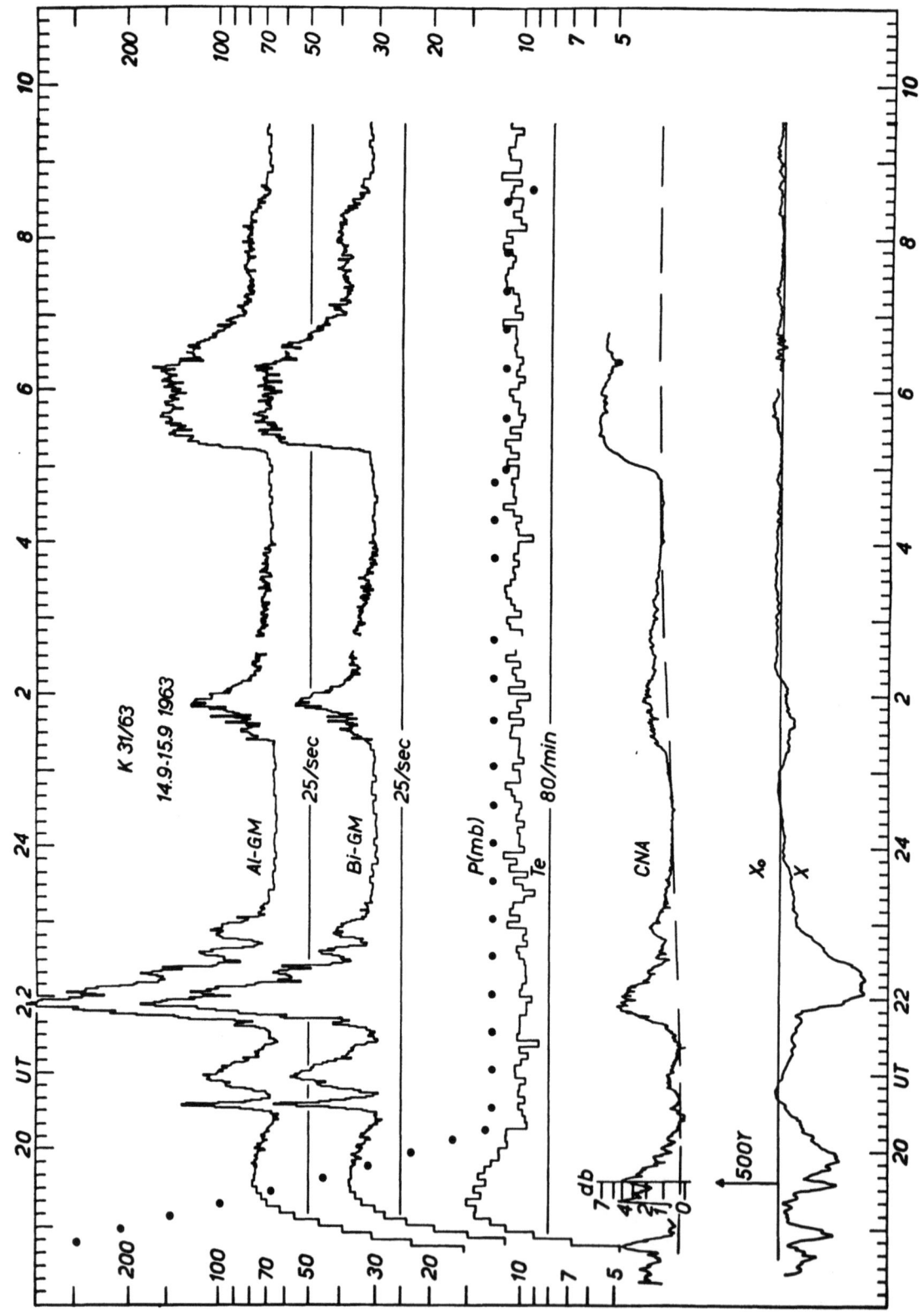

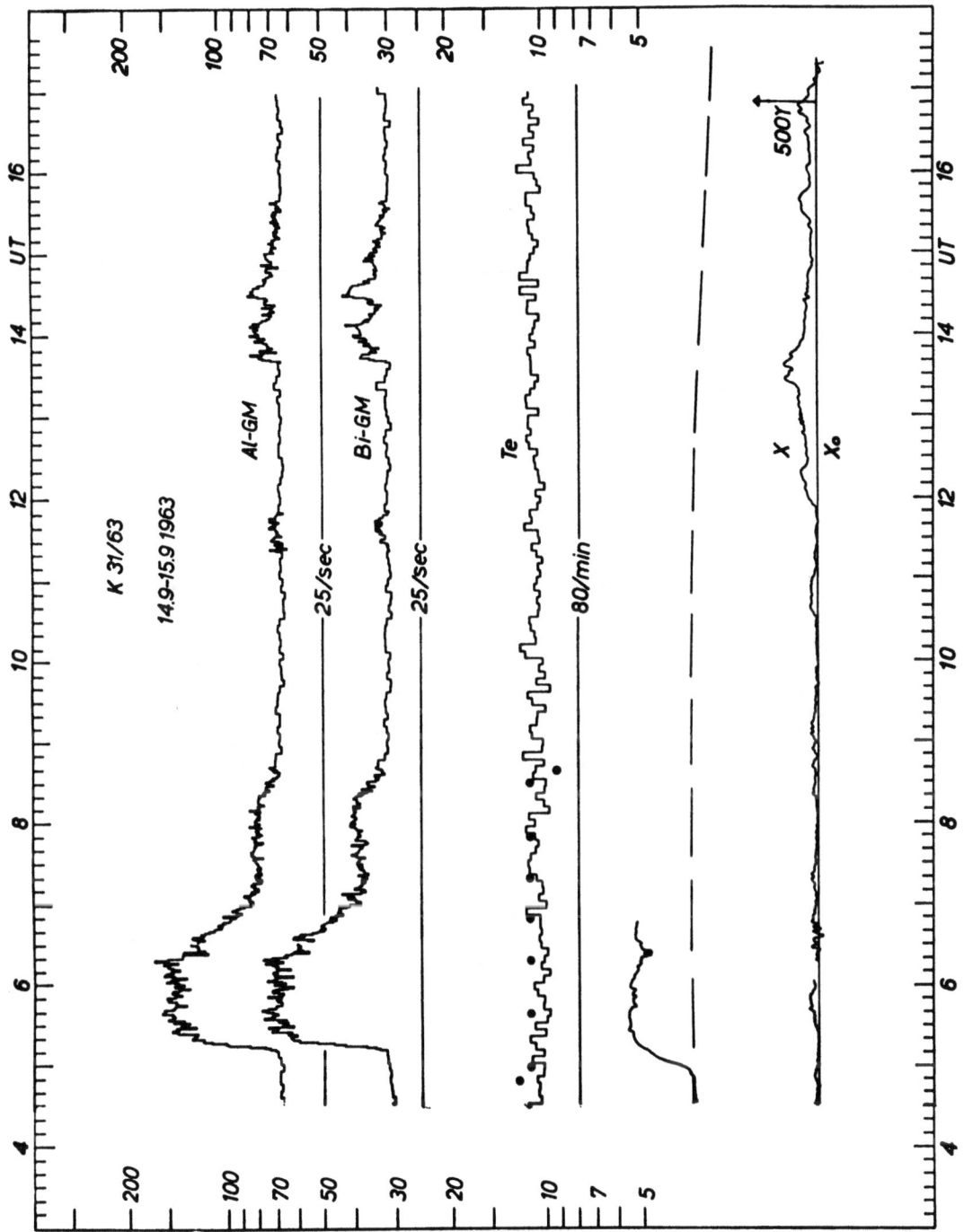

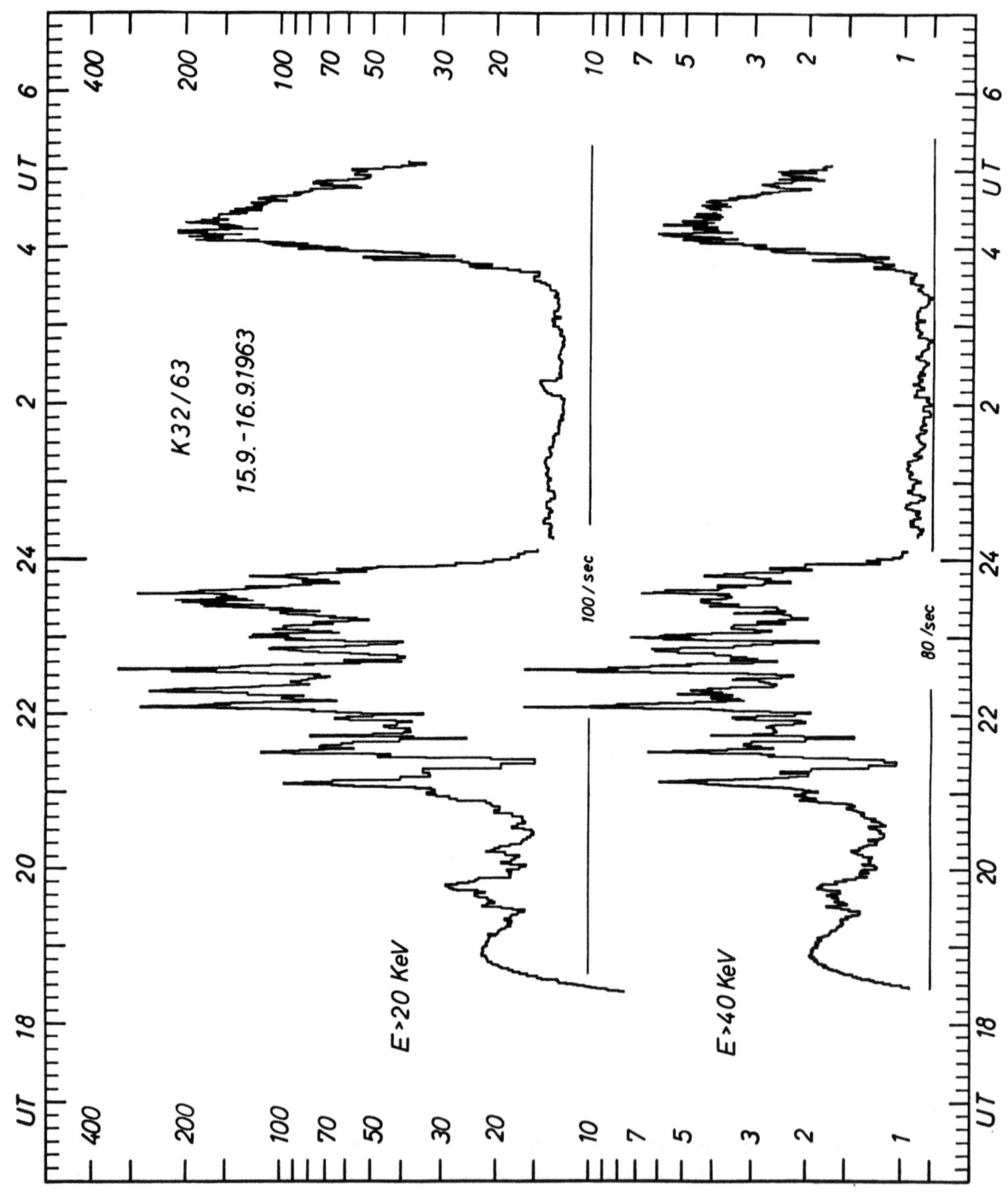

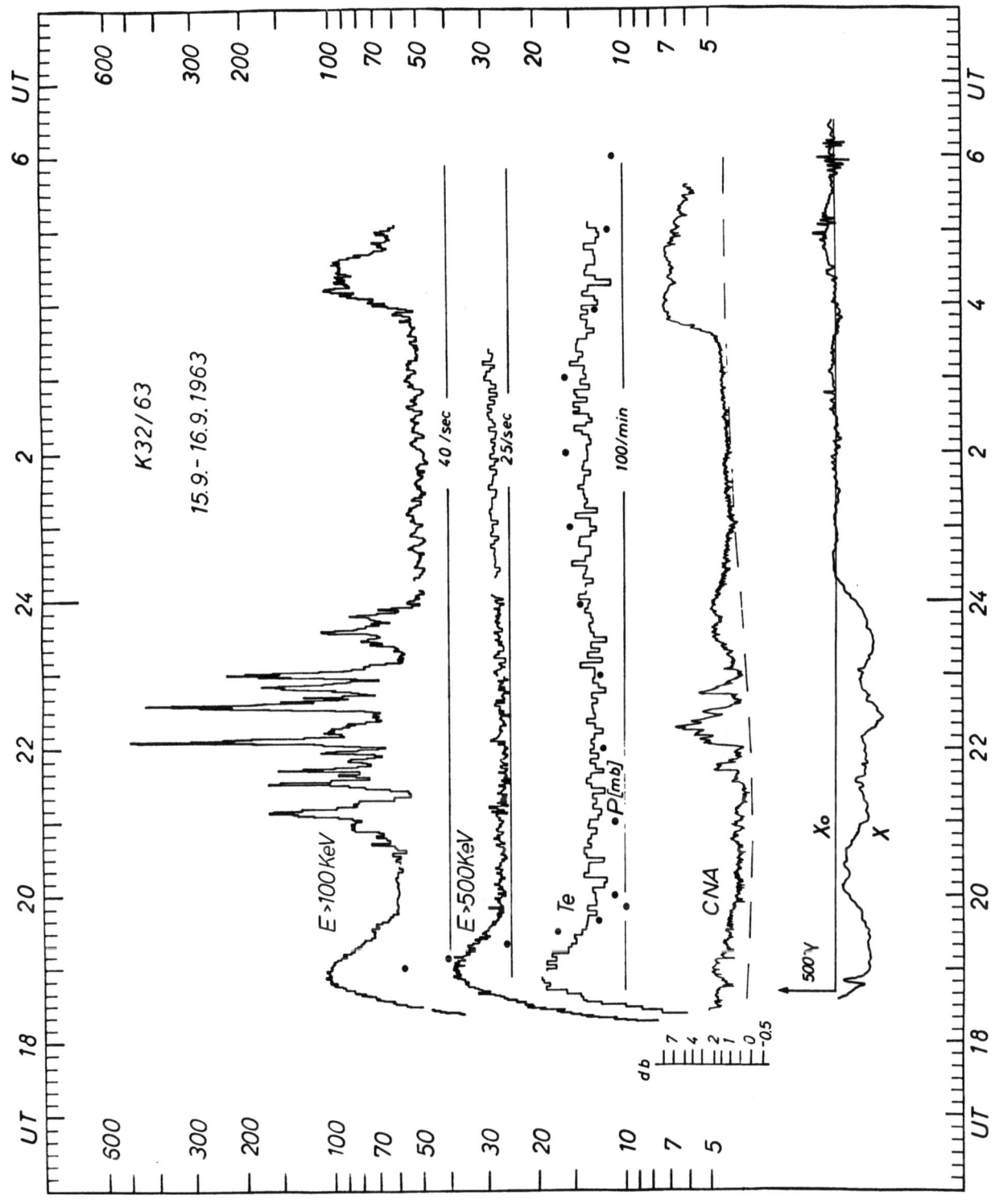

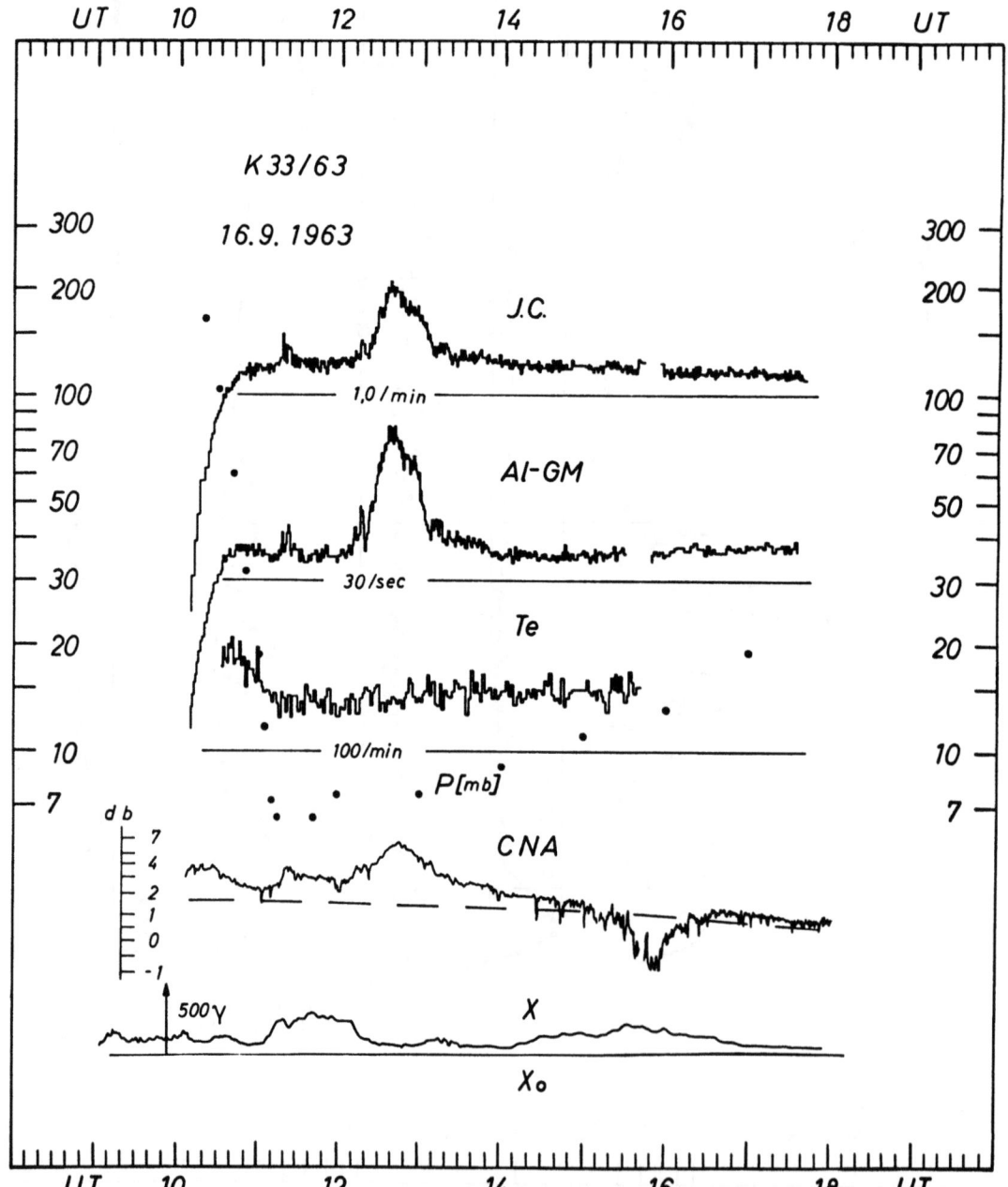

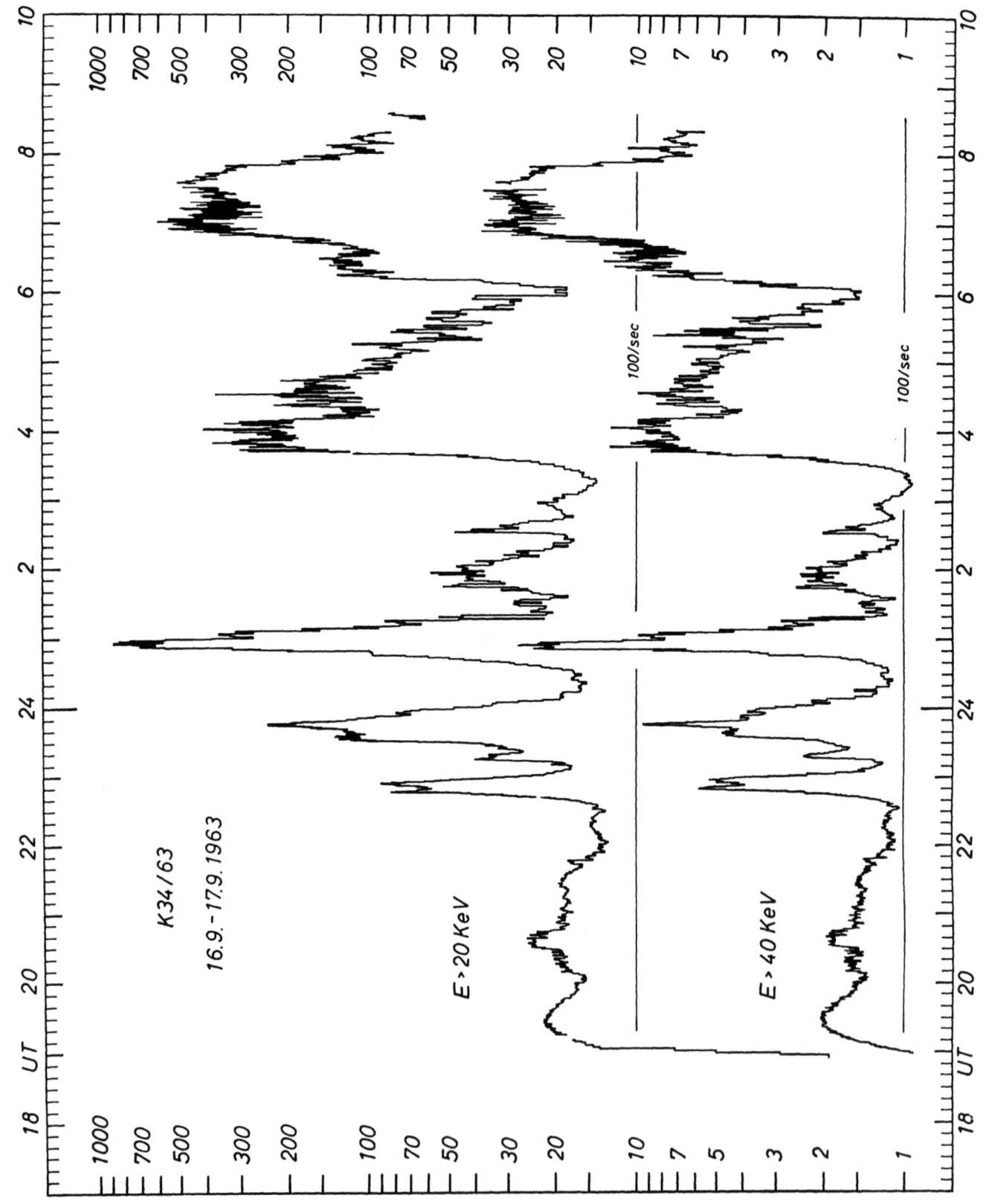

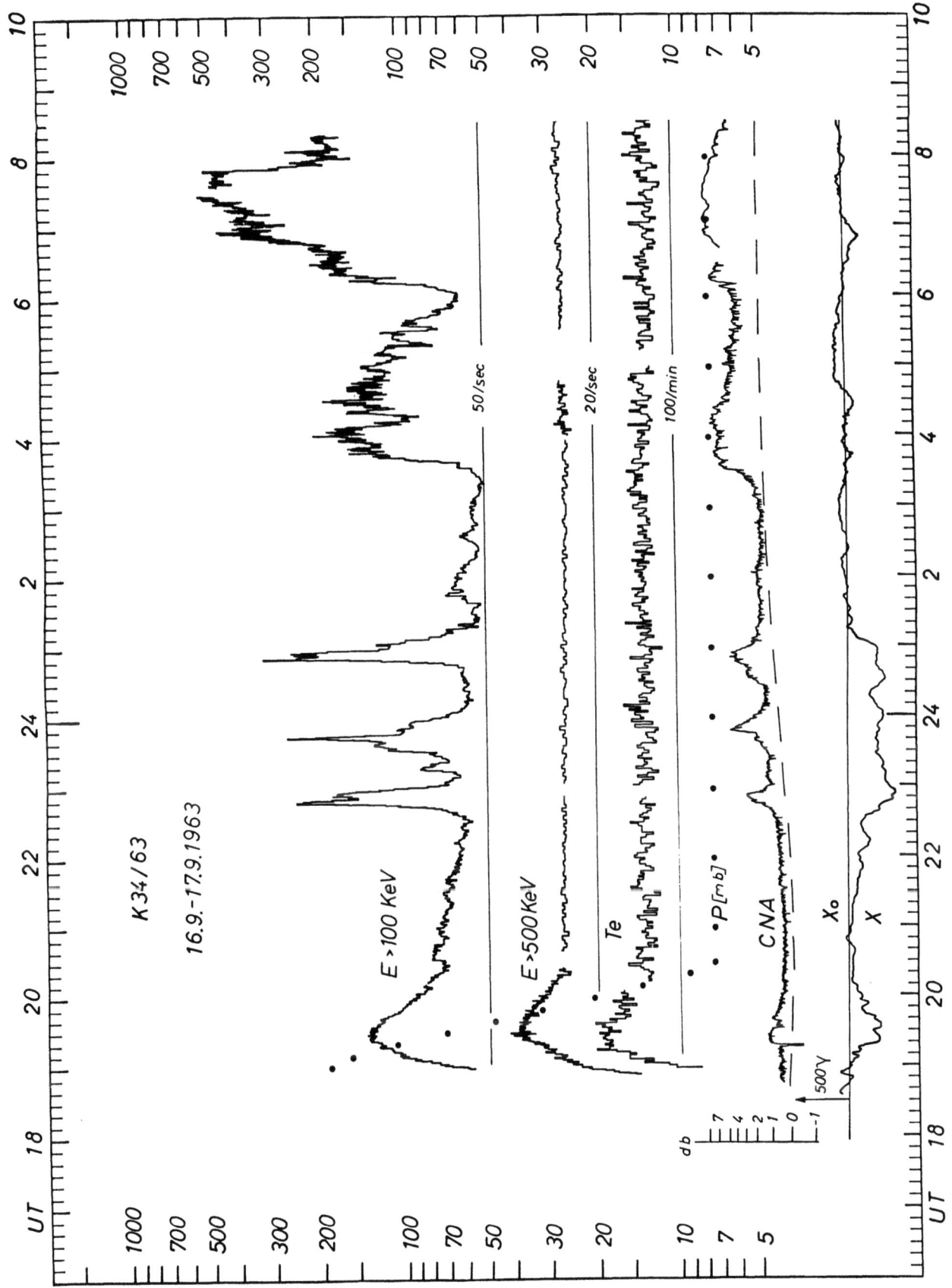

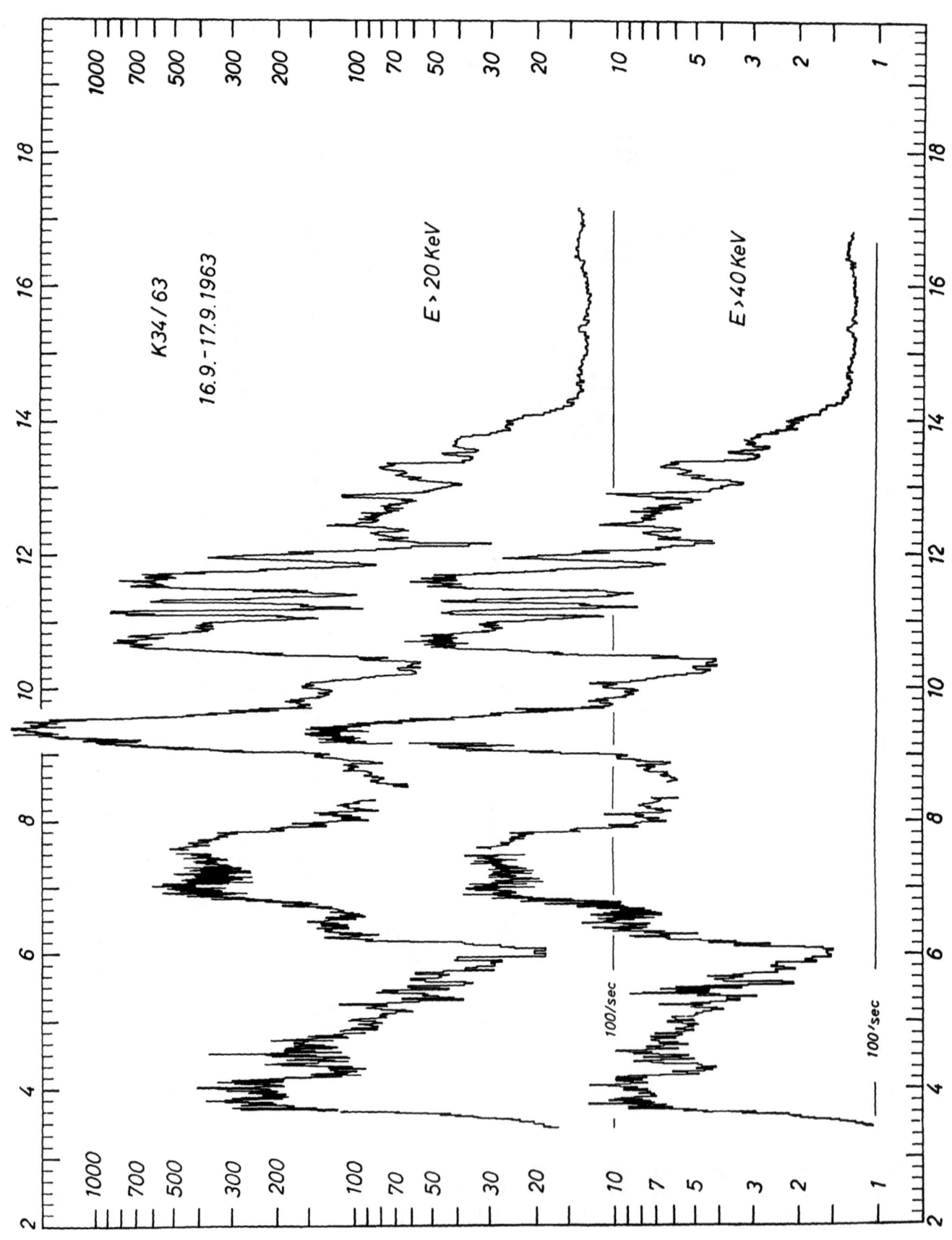

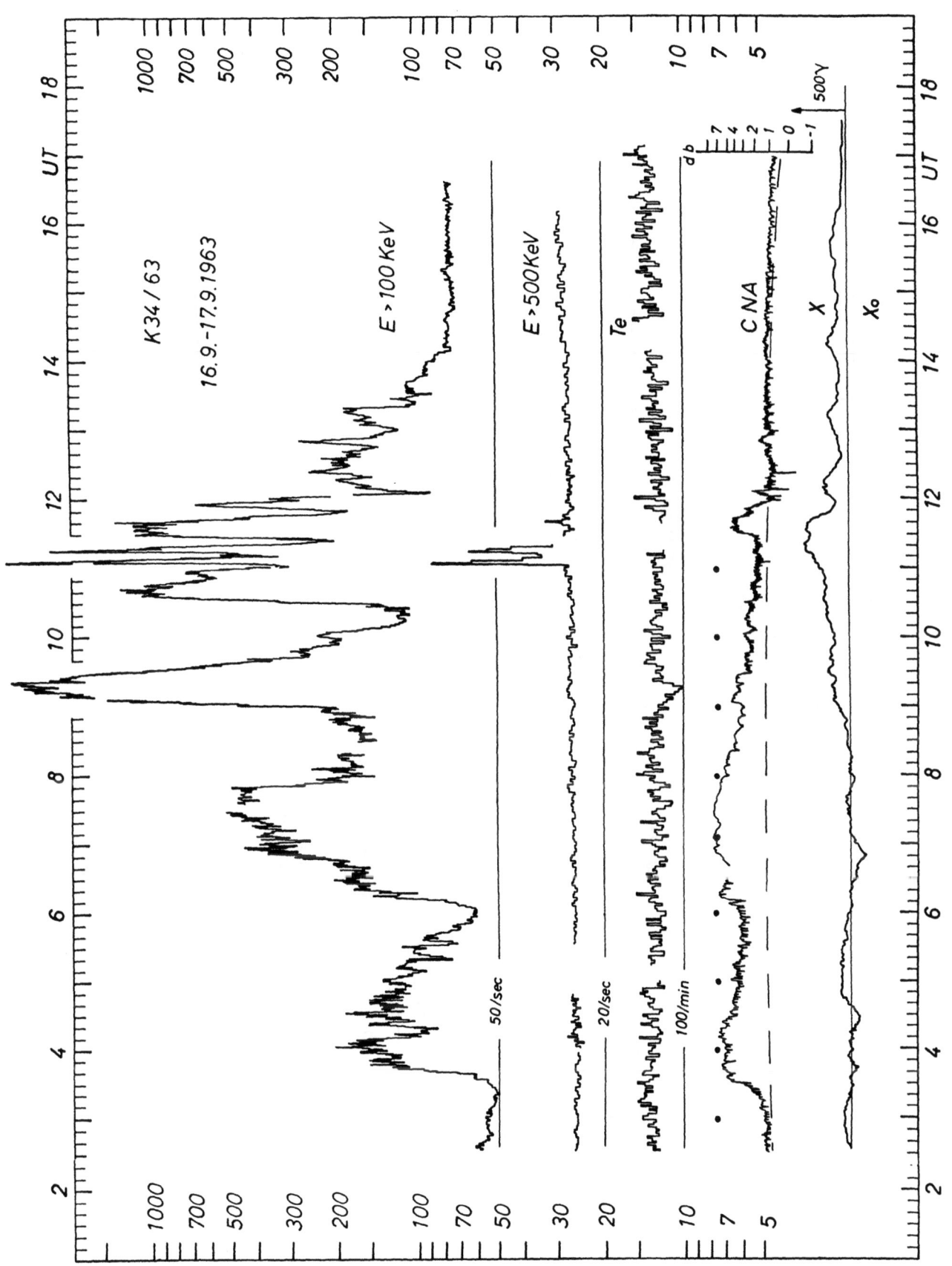

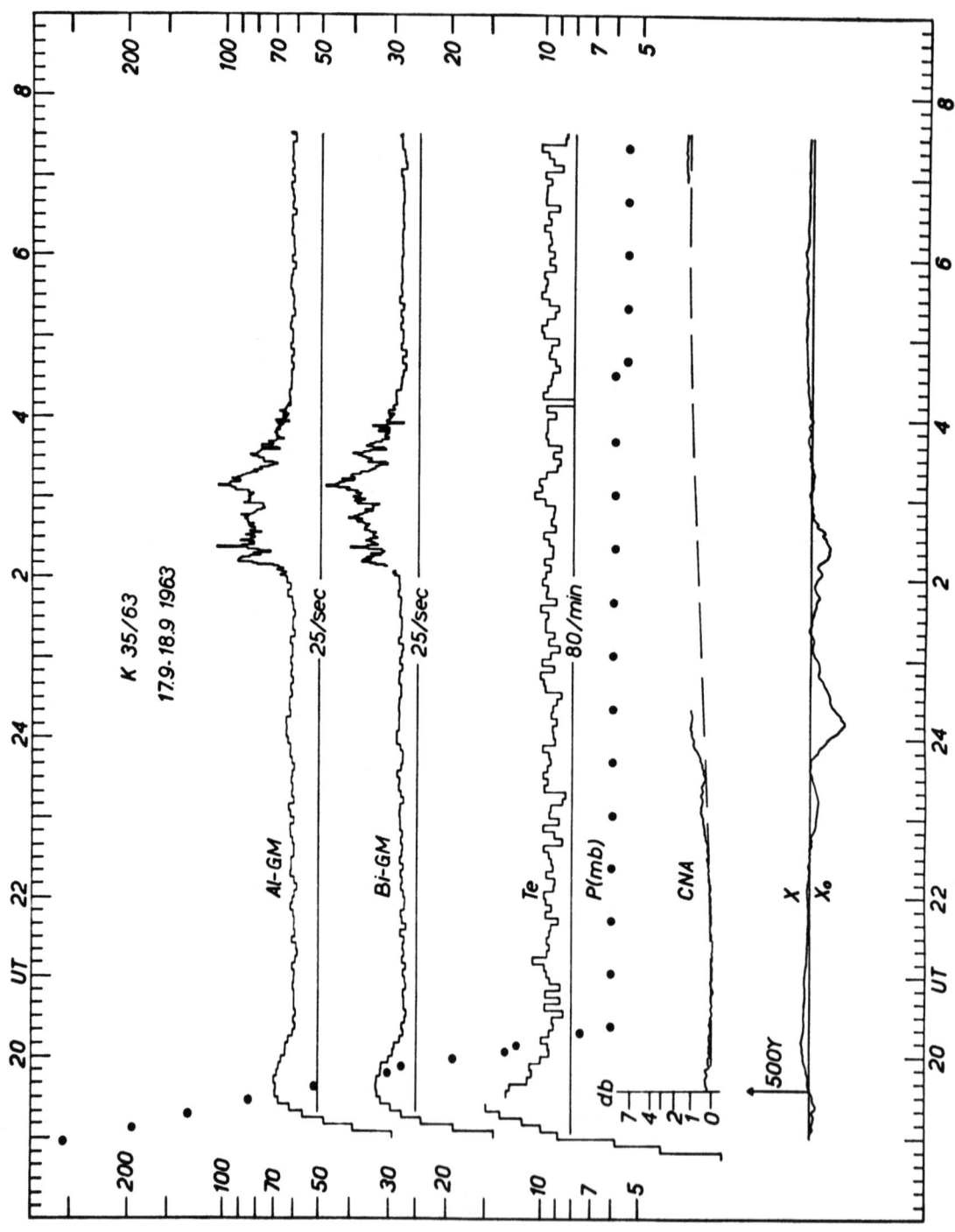

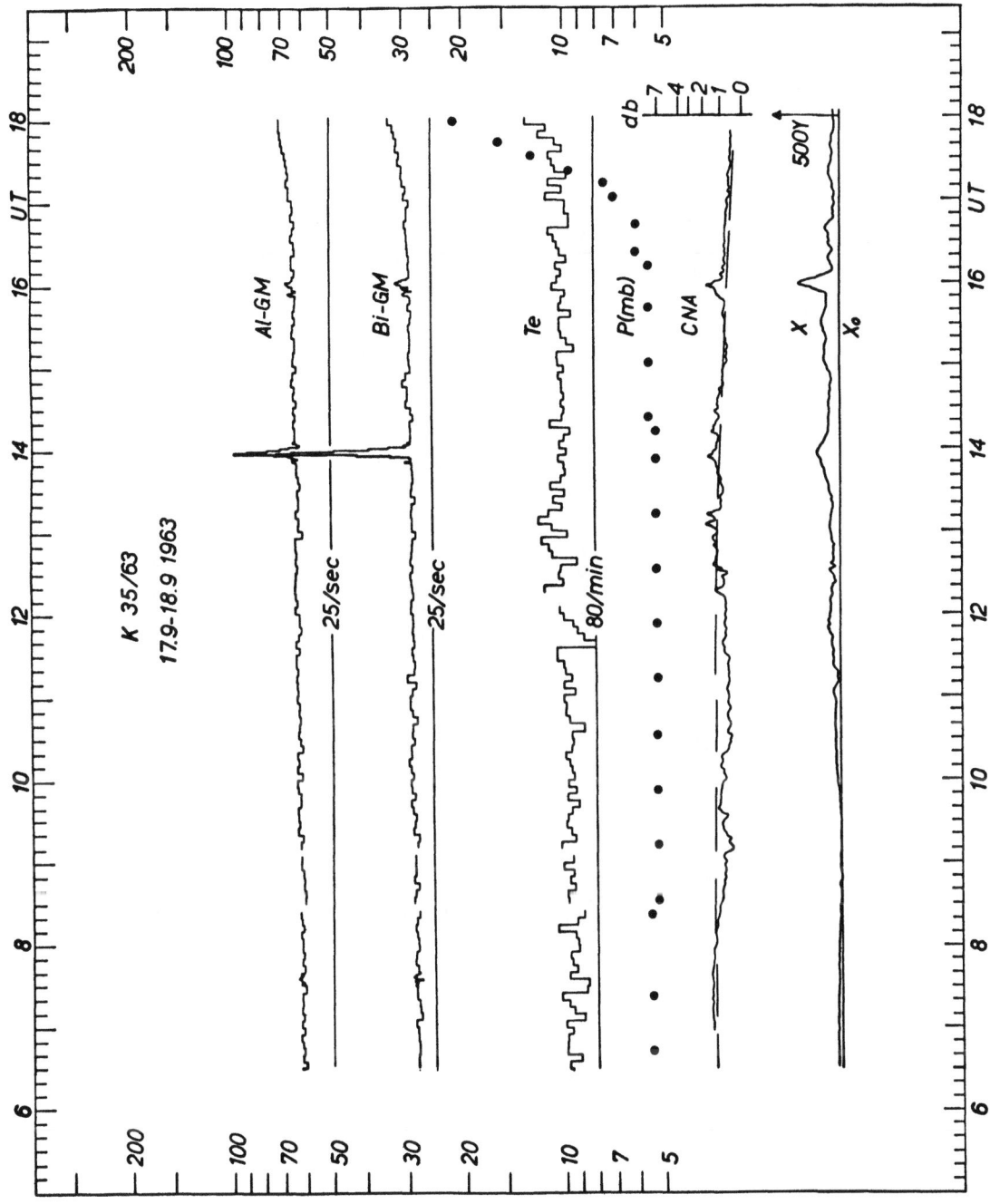

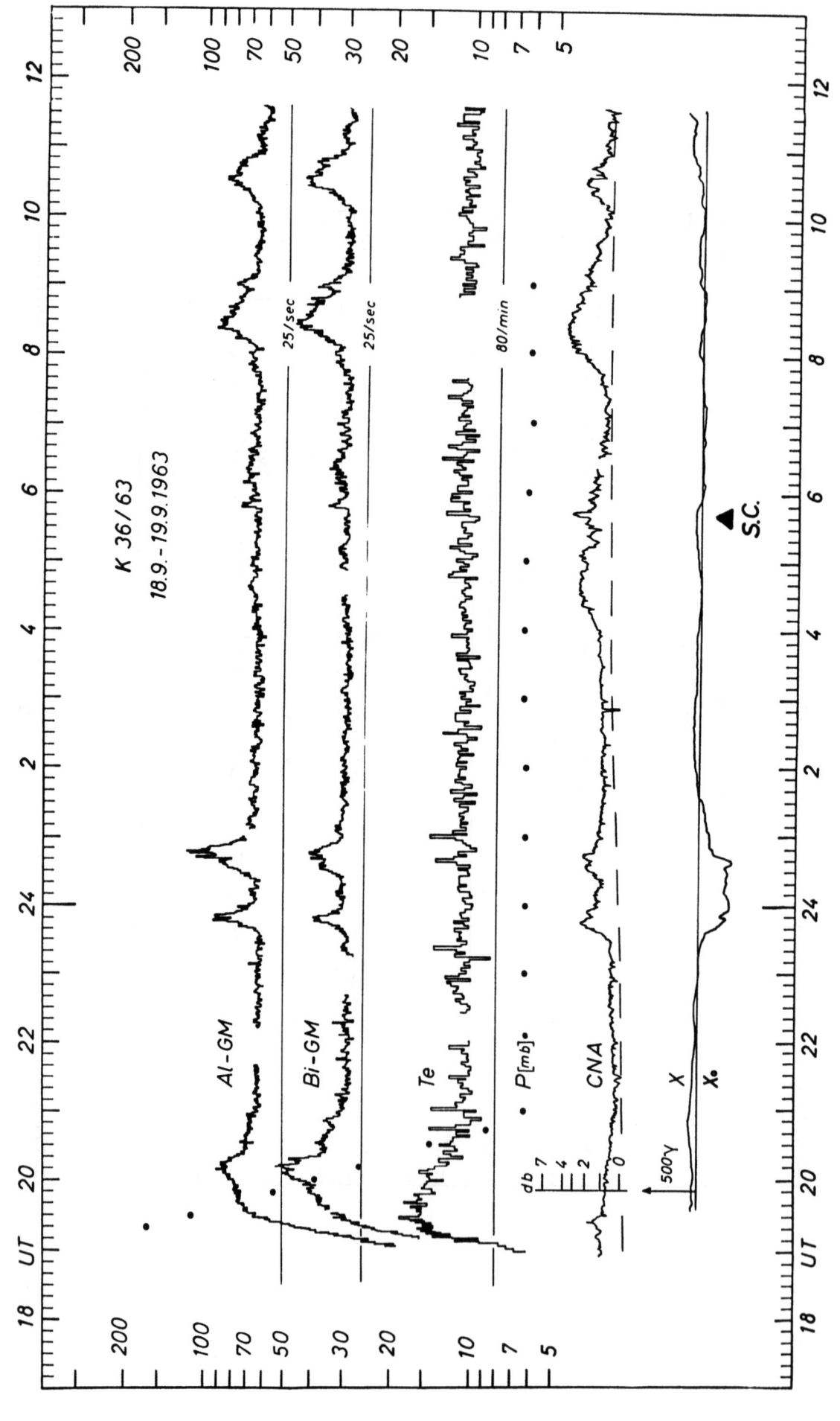

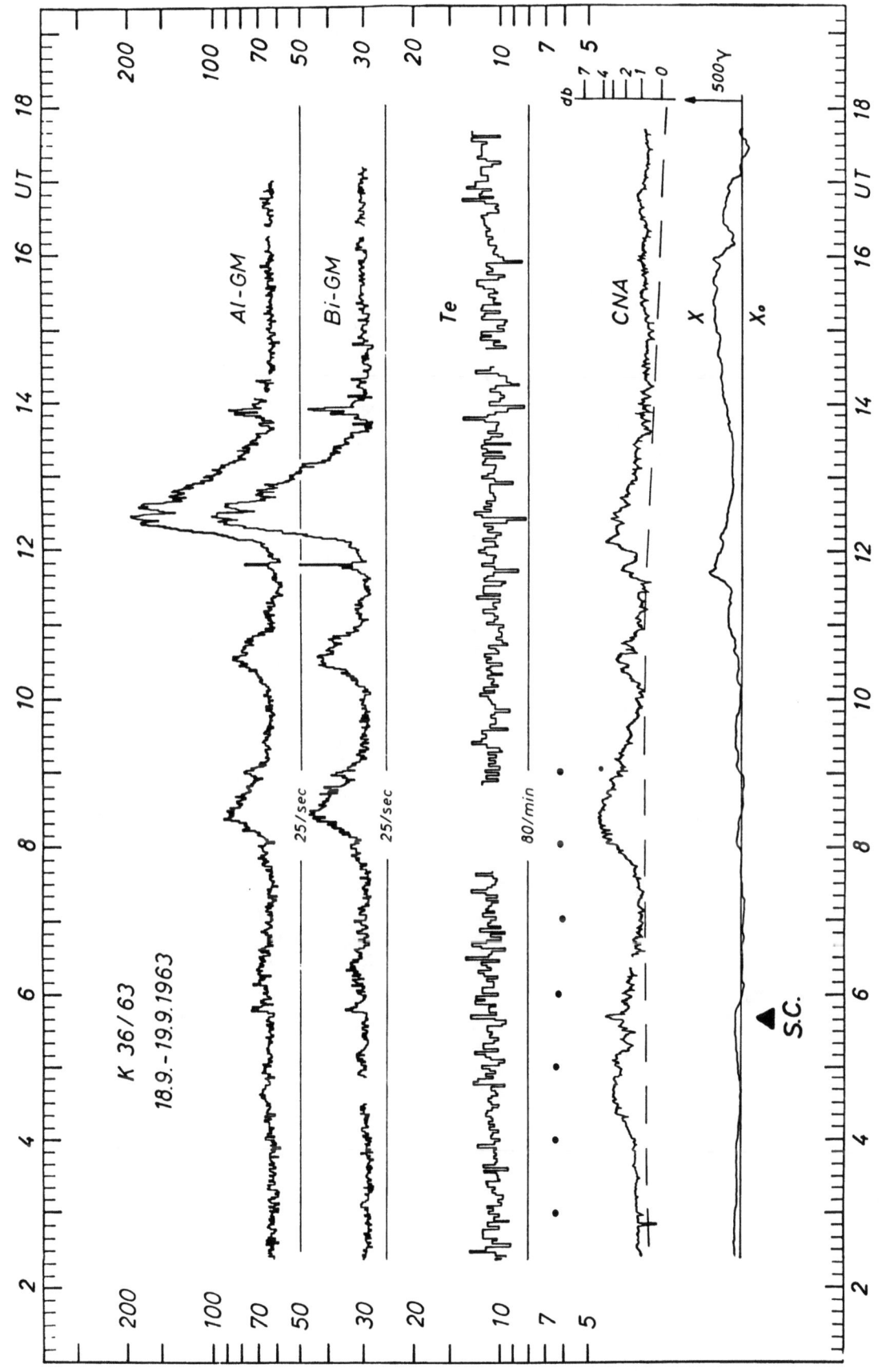

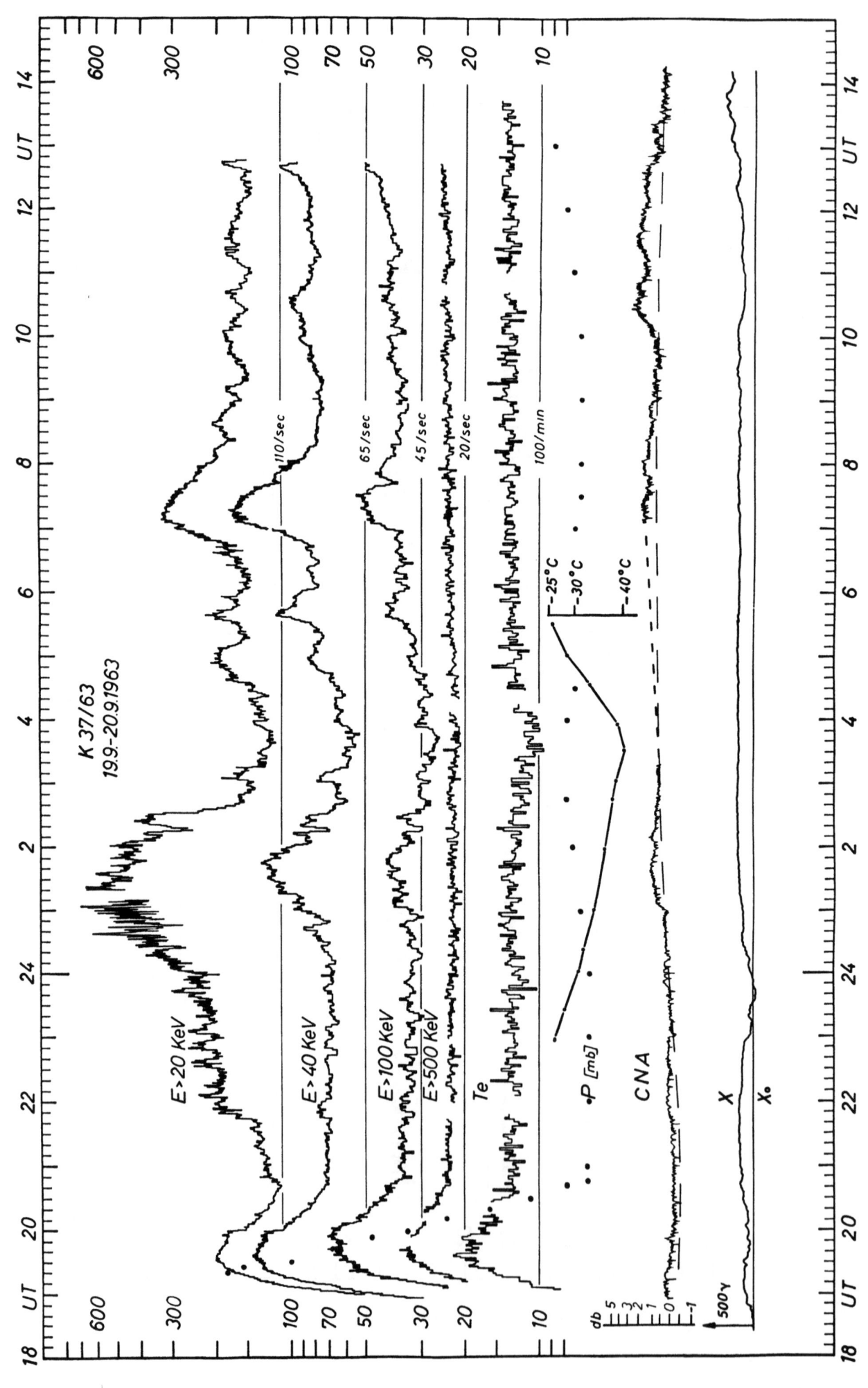

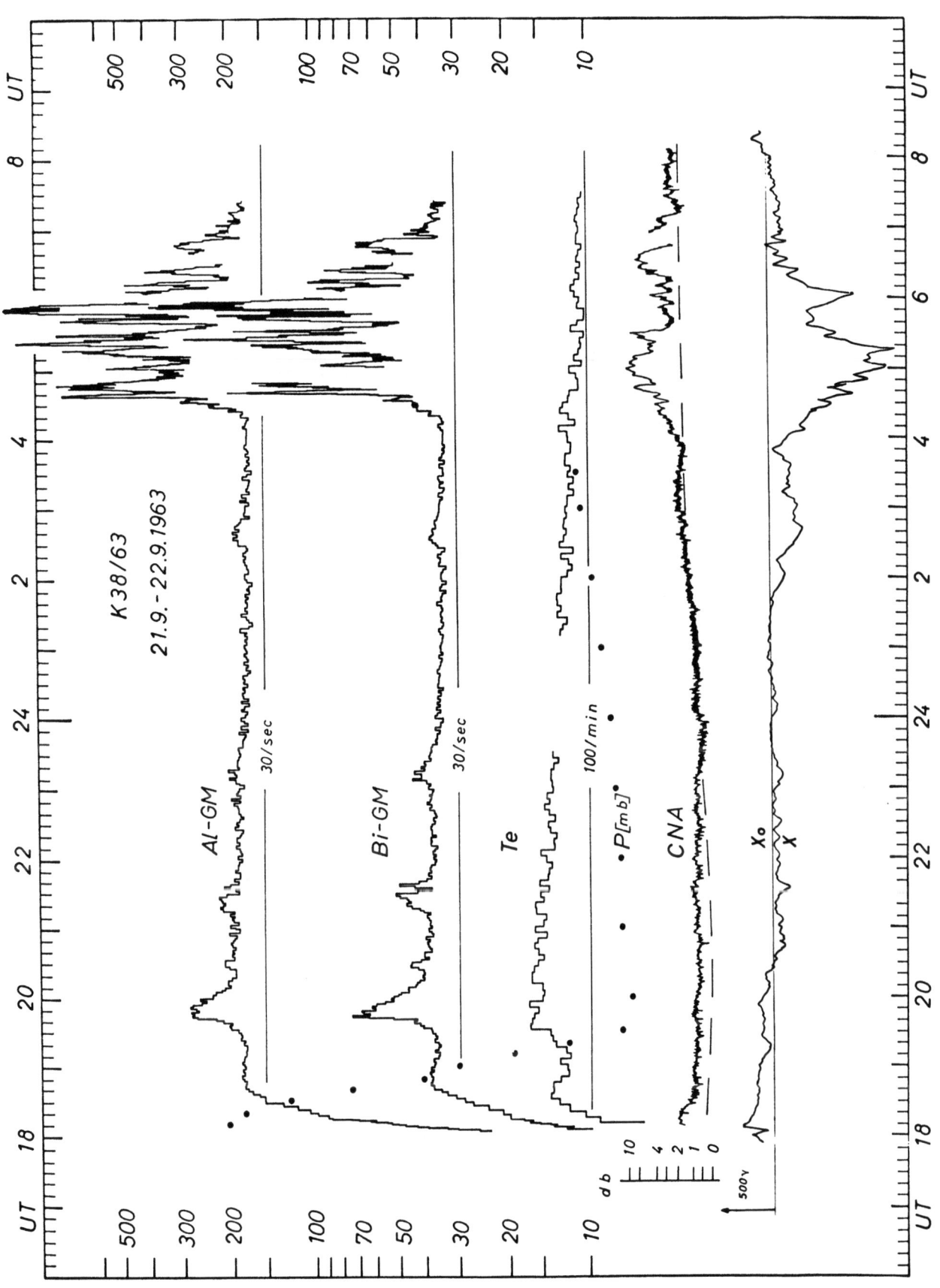

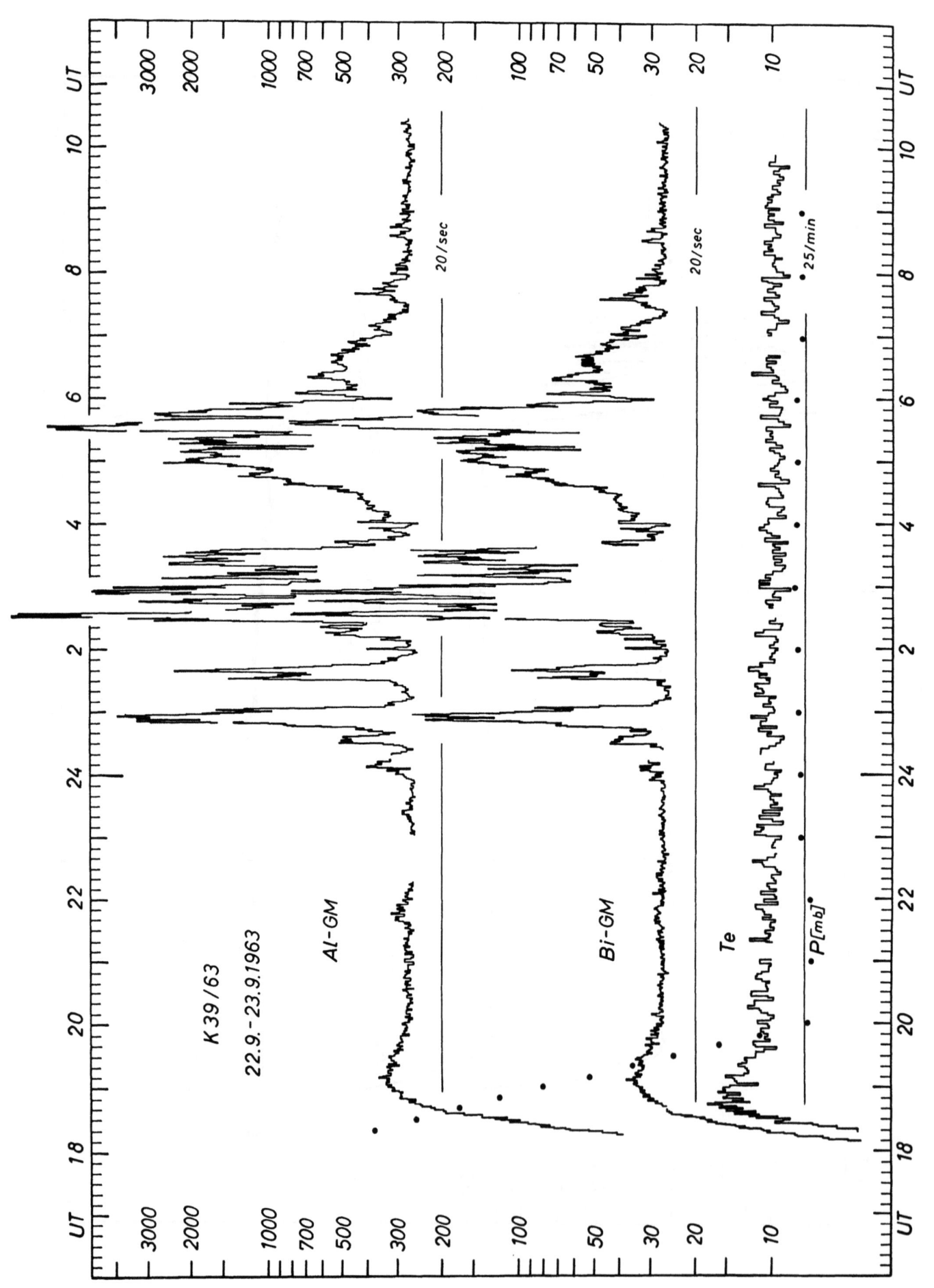

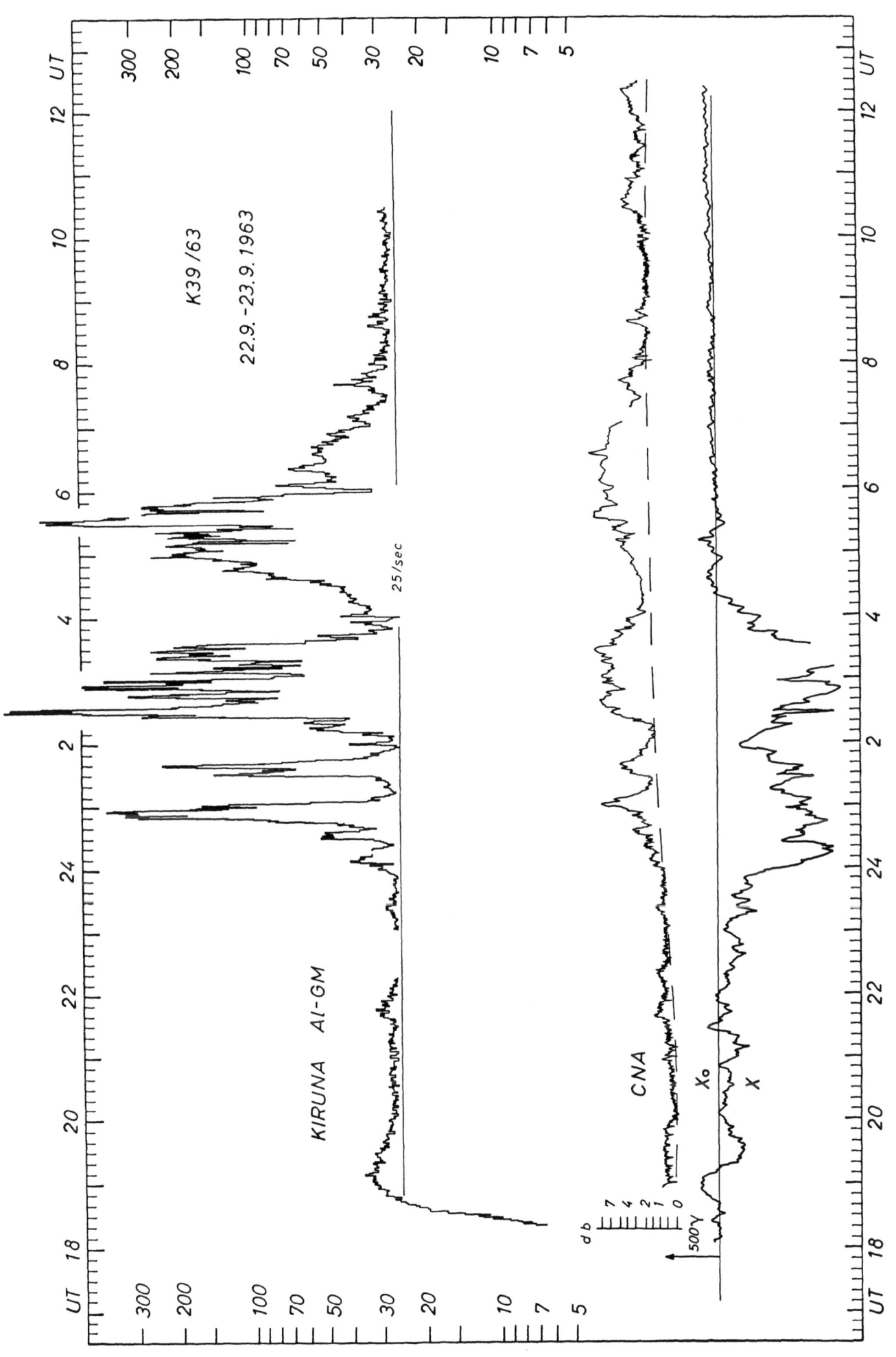

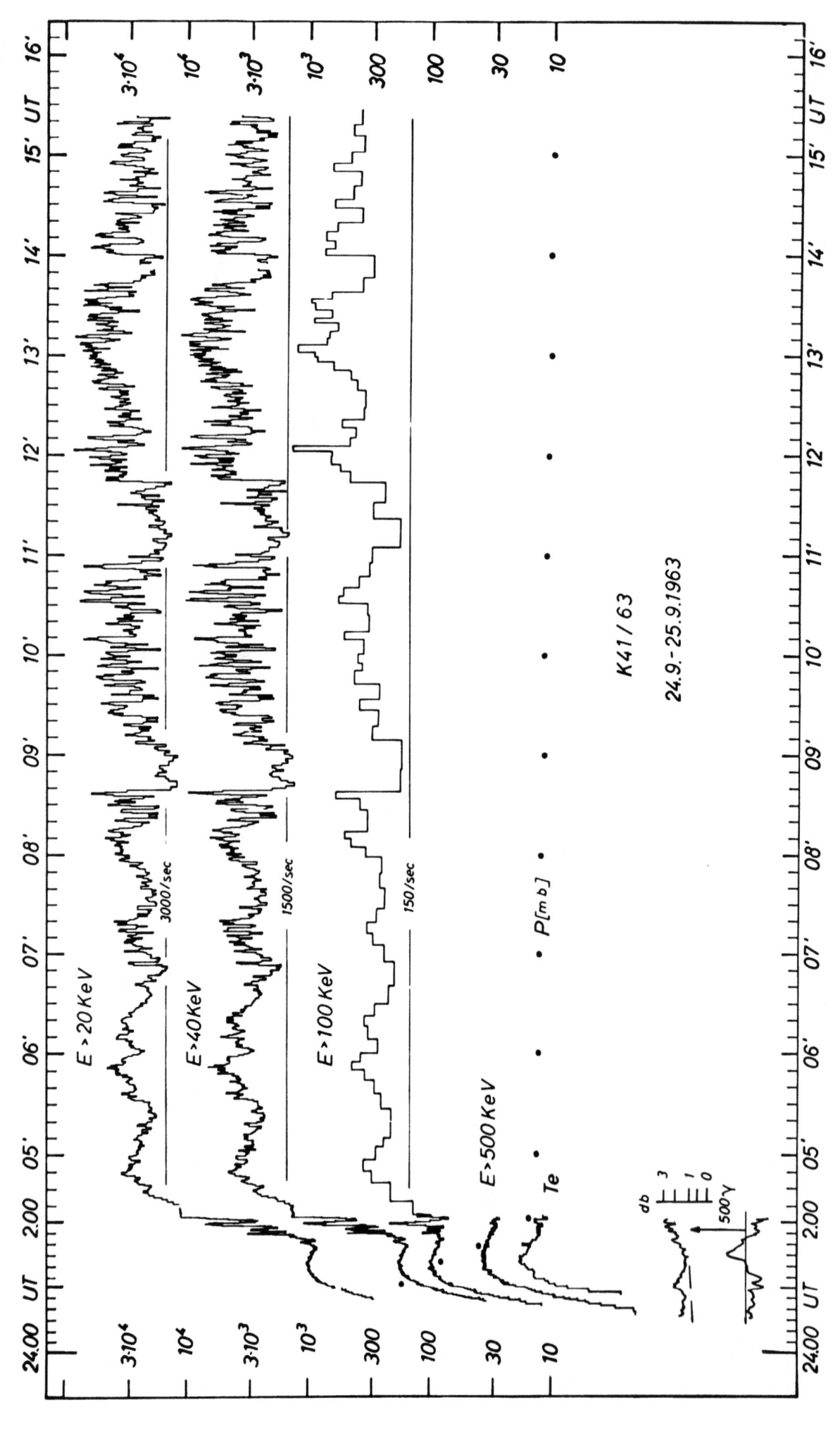

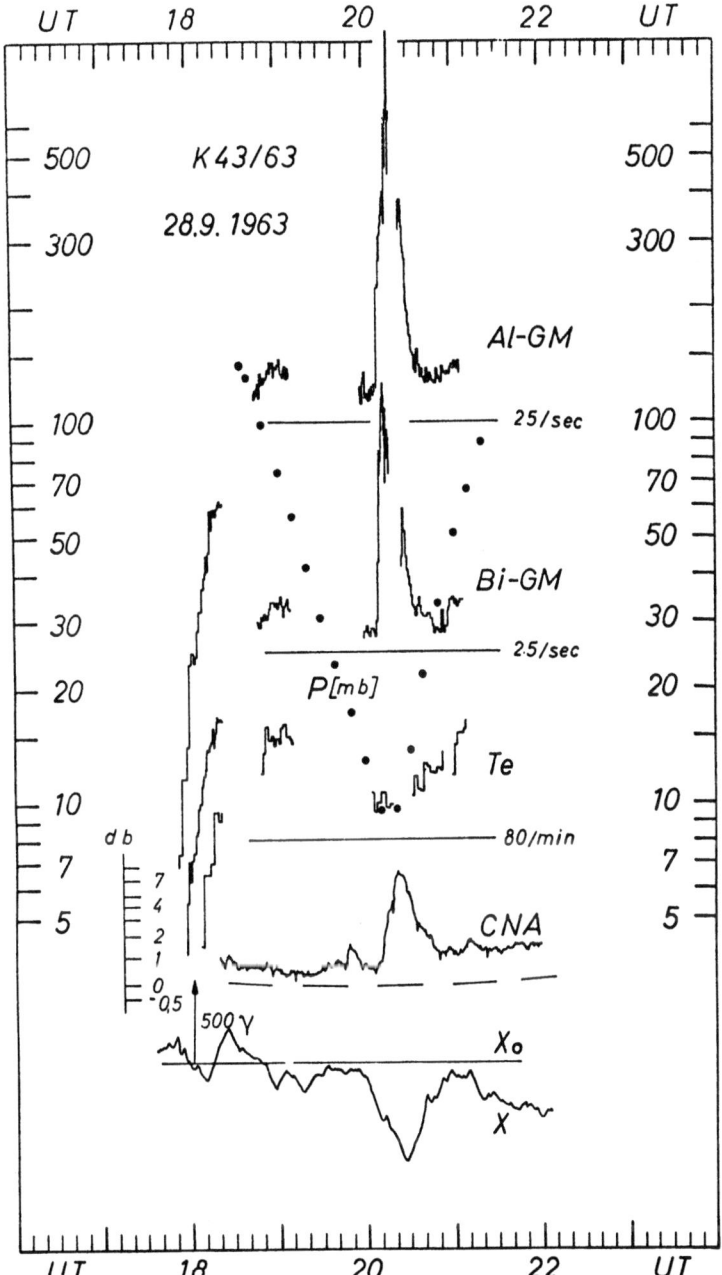

If you have any concerns about our products,
you can contact us on
**ProductSafety@springernature.com**

In case Publisher is established outside the EU,
the EU authorized representative is:
**Springer Nature Customer Service Center GmbH
Europaplatz 3, 69115 Heidelberg, Germany**

Printed by Libri Plureos GmbH
in Hamburg, Germany